Survey of Industrial Manipulation Technologies for Autonomous Assembly Applications

NIST-Internal Report #7844

Roger Bostelman, Joe Falco

Intelligent Systems Division
National Institute of Standards and Technology
Gaithersburg, MD 20899
roger.bostelman, joe.falco @nist.gov

February 29, 2012

Disclaimer: NIST does not endorse products discussed within this report nor manufacturers of these products. Products mentioned are for information purposes only and are not expressed as an endorsement for them or their manufacturer.

Table of Contents

Executive Summary 3
1. Introduction 4
   a. Background 4
   b. End Effectors for Assembly 5
   c. Survey Objective 7
   d. NIST Mobility and Manipulation Project 7
2. Current Robotic Assembly Systems 8
   a. Gripper Types 9
      i. Class I – End-effectors with fixed shape fingers 9
      ii. Class II - Special purpose end-effectors 11
      iii. Class III - Multipurpose end-effectors 14
   b. End-Effector Sensing 16
      i. Tactile 16
      ii. Proximity 18
      iii. Force/Torque 19
      iv. Vision 20
   c. Generic Robot System Assembly Tasks 22
      i. Insertion 22
      ii. Bin Picking 24
   d. Specific Robot System Application Examples 25
3. Advances in Robotic Assembly 26
   a. Systems 26
   b. Research 28
      i. Touch 28
      ii. Force 29
      iii. Vision 30
      iv. Threaded Insertion 30
      v. Bin Picking 31
      vi. Control 31
      vii. U.S. Government Research 33
   c. Patents 35
4. Summary 37
5. Recommendations 38
6. References 41
7. Appendix 47
   a. Terminology 47
   b. Industrial Robot System Standards 57
   c. Design for Assembly – Boothroyd-Dewhurst Method 59
   d. Types of Force Control Functions 60

Executive Summary

Robots are often perceived as fast, precision machines for replacing humans. However, equating the abilities of robots and human beings is risky business. A task that appears easy for a human assembler can be difficult or even impossible for a robot. In today's manufacturing environments, to ensure success with robotic assembly, engineers must adapt their parts, products, and processes to the unique requirements of the robot.

This document provides an overview of the current state of manipulation systems, as well as insight into future manipulation systems through discussion of research being performed in the field. The objective of this survey is to educate the manufacturing community about industrial manipulation capabilities, advancements, and research while focusing on autonomous assembly applications. The paper addresses advanced dexterous manipulation by surveying the state-of-the-art in end-effector and manipulation capabilities for autonomous assembly tasks.

Humans can learn complex assembly tasks with relative ease compared to a robotic learning system. While today's robots are precise and capable of achieving repeatable positions to within a few thousandths of an inch over thousands of hours of operation, dimensions and surfaces can vary between parts. If these tolerances between like parts are significant, a conventional robotic system using a position-based robotic control strategy based on nominal part dimensions will be ineffective as an assembly tool in cases where the assembly tolerance is less than the positional uncertainty of the robot. In these cases, end-effector sensing and an adaptive control scheme must be implemented in order to meet the requirements of the assembly operation.

Results of this survey show that dexterity, perception, and force control of robot systems are advancing as exemplified by the availability of multi-fingered and tactile grippers, random part identification and localizing capabilities, and research aimed at advanced tasks such as assembling parts while in motion. Advanced military programs are underway to improve the state of science for robot dexterity and capability. Safety and performance standards are beginning to consider collaboration among humans and robots within the same workspaces. On the manufacturing forefront, advances in highly capable assembly systems are minimal and recent gripper patents are mainly special purpose. Following is a set of generic recommendations for industry actions and new technologies towards the development of more capable robotic systems:

Actions:
- Development of performance measures for assembly
- Increased use of robots for 'intelligent' fixturing
- Development of methodologies for human/robot interaction
- Integration of current industrial grippers with sensors
- Development of autonomous robot systems capabilities that compare to humans
- Definition of clear, standard interfaces for grippers
- Use of robot hand guiding through the teaching process
- Development of path planning for two arm robots
- Use of external metrology to support robot system applications
- Verification of perception systems combined with low cost gripper tactile sensors
- Improvement of force control for assembly processes

Technologies:
- Adaptive end-effectors
- More compact end-effectors and end-of-arm sensing
- Tactile sensing for low cost pneumatic grippers
- Robots on vehicles
- Dynamic robot work volumes
- Tactile based response for robots

- Collaborative robots

1. Introduction

a. Background

"The logical evolution of the assembly line would seem to lead to one that is fully automated. Such an automated system would ideally imply the elimination of the human element and its replacement with automatic controls that guarantee a level of accuracy and quality that is beyond human skills. In the 1980s, Japanese and Italian automobile manufacturers so successfully automated their assembly lines that certain of their factories consisted almost entirely of robots regularly doing their jobs. On the other hand, General Motors found that robots could not replace human workers and had to retrench from technology and focus on retraining workers." [Bruno, 2011]

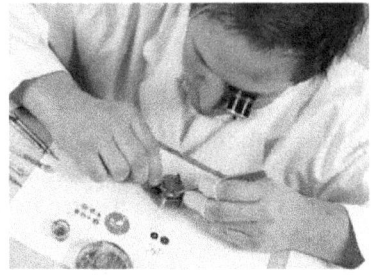

[Genaldy, 1990] et. al., compared human and robot performances for a simple assembly operation through a time and cost study finding that a pair of robots performing the same simple task were slower than the human. Despite the significant research that has been performed in robotic manipulation and autonomous assembly, it is evident that the technology has a long way to go to match the assembly capabilities of, for example a 16th century mechanical watchmaker including his/her: dexterity, sensitivity, accuracy, and compliance. (*Watchmaker photo copyright approved: Igor Gratzer/Shutterstock*) However, great strides towards this capability have been made for at least larger automotive part assembly as stated in [Picard, 2002] with watchmaker precision. Their FlexPlace system achieved sub-millimeter accuracy and did away with traditional heavy and expensive tooling by using improved robot path adjustment through learning, part modeling/matching, and a set of sensors mounted on the robot gripper.

The goal of this document is to provide an overview of the current state of manipulation systems as well as insight into future manipulation systems through discussion of research being performed in the field. This study is not exhaustive and instead an example of the current state of the art. Discussion will focus on the end effector (mainly robot grippers) and the combined end-effector/manipulator capabilities to perform autonomous assembly tasks. The need for agile manufacturing is moving the manufacturing industry towards more flexible and complex assembly operations involving, for example:
- multiple manipulators providing flexible part manipulation and fixturing;
- assembling parts while the manipulator and/or the parts are in motion;
- and human/robot collaborative assembly where humans and robots work "hand-in-hand" to assemble a common and complex system of parts.

Manipulators can have varied end-effectors ranging from basic parallel gripping of a particular part to complex dexterous hands capable of grasping parts of various geometries. Gripping/grasping and manipulation combined with perception of the part, manipulator, and environment further complicates the potential tool set that may be required for generic autonomous assembly across the variety of manufacturing applications including for example: automotive, aircraft, distribution, and food-processing.

Selecting an end-effector to perform a manipulation task requires an analysis of the parts to be grasped. In the simplest case, as in many industrial robot applications, only a few parts must be handled through a predefined set of tasks. In the case of a known part and known grip, an end-effector system must be defined to grasp and maintain control of a part through the various manufacturing tasks. Known part – known grip applications use fixturing to constrain parts in known locations. In another case, called known part, the parts are known, but there is variation in part pickup orientations (such as in random bin picking applications) and a grasping algorithm must be applied to determine the best grasp to use in order to access a part and maintain control. The last case, called unknown part, is the least structured environment where the robotic system must both determine the identity of a part and its properties (e.g., shape, size, mass, position) and subsequently a valid grasp. The more unstructured the environment, the more advanced the supporting perception system must be. Ideally, but currently unrealistic, is a robotic manipulation system having the dexterity and perception of a human to work in an unstructured manufacturing environment.

b. End Effectors for Assembly

End-effector designs are quite diverse. The flexibility of end-effectors to handle a wide range of objects without manipulator mechanical reconfiguration leads towards the use of a universal robotic hand. These universal robot hands, developed to mimic multiple human grasp techniques, require very complex mechanisms and control algorithms. They are an extensive area of research, but to date no practical, commercially-available versions exist. More economical end-effectors, designed using a single grasp technique to handle a subset of objects of varying weight, shape, and material are often termed grippers. These grippers are typically designed with two fingers having single degree-of-freedom (DOF) using various kinematic mechanisms and specialized jaws for the prehension of a particular object or family of objects. Sometimes grippers may incorporate three single DOF fingers for grasping object geometry without the need for specialized jaws (e.g., objects with cylindrical features). Some end-effectors are designed for even more specialized prehension operations. A classification scheme for end-effectors [Nof, 1999] is outlined in section 2a Gripper Types with examples of each current end-effector type for use in robotic assembly.

[Monkman, 2007] states that the choice of gripper depends mainly on the work it has to perform and that every prehension task is characterized by the following factors and requirements:
- Technological requirements: prehension time, gripping path, time dependence of the prehension force, and the number of the object acquisitions per gripping cycle
- Effects of the prehended objects: mass, design, dimensions, tolerances, position of the center of gravity, stability, surface, material, strength, and temperature
- Factors related to handling equipment: positional accuracy, axial accelerations, and connection specifications (mechanical, electrical, fluidic, etc.)
- Factors related to environmental parameters: process forces, feeding conditions and clamps, storage conditions, contaminations, humidity, and vibration

[Monkman, 2007] also suggests that gripping procedures consist of part prehension and retention and can be divided into four phases:
- Preparation for contact e.g., by appropriate orientation of objects following a predefined motional pattern.
- Prehension by establishing contact between object and gripping surfaces. At this stage the work piece is subjected to static forces and moments.
- Retention of the part during its manipulation in space or, in some cases, moving, rotating, or even (in rare cases) mounting. Dynamic forces and moments occur in the course of motion or task related procedures.
- Release of the part at its destination, e.g., by switching-off the vacuum supply and possibly using the assistance of an integrated ejection mechanism.

How well a part is secured in a gripping process depends on the number of degrees of freedom allowable following prehension. Figure 1, modeled by the National Institute of Standards and Technology (NIST), shows Monkman's typical two-point prehension on various parts and the corresponding remaining degrees of freedom.

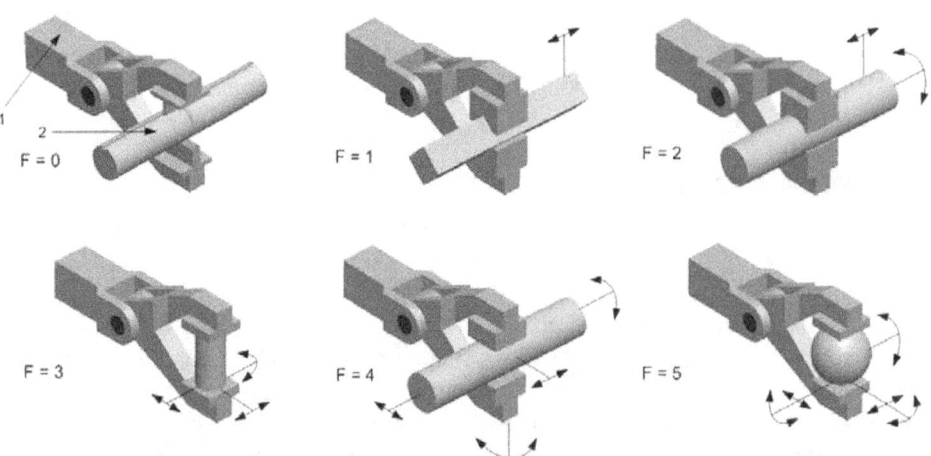

Figure 1 – Two point prehension of parts showing the degrees of freedom F = 0 to F = 5.
(1 gripper jaw, 2 part)

Monkman says that a part is held by force matching (with frictional or gravitational forces equal to the gripping force). Clamping forces from pinching grippers cannot be arbitrarily increased. The upper limit force is dictated by the allowable surface pressure, depending in turn on the contact force and the coefficients of elasticity of the gripper jaw and part material. Imperfections in gripper finger form can lead to poor surface contact with the part. This becomes particularly apparent when long fingered grippers are used. The design of the gripper system is influenced significantly by the forces necessary to ensure reliable prehension of the part. However, the required gripping force depends on many factors which can be only partially estimated. Some of these factors are:
- Spatial settings: arrangement of the gripper relative to the industrial robot and its movable axes
- Resultant force: vector sum of all single acting forces resulting from mass, inertia, Coriolis, and centrifugal forces – all of which may change with robot movement
- Geometry of the part and prehension points

The ultimate part retention stability is achieved by maximum matching of the gripper and part profiles. Jointed grasping mechanisms also make it possible to compensate for irregular part shapes and to correct for position deviations. Sensitive/delicate parts should be handled with shape matching having no appreciable impactive forces.

The strategy for gripping a part can be:
- Predetermined: The operations required to achieve a reliable grip at the corresponding contact points are pre-programmed
- Variable: operations are only briefly defined and can be adaptively matched to the situation in accordance with the information supplied by sensors

Impactive gripping (impact of jaws against part surfaces) requires the motion of solid jaws in order to produce the necessary grasping force. Ingressive gripping results in surface deformation or even penetration (intrusive) of the surface down to some predefined depth (force-shape mating). Contigutive prehension implies a direct contact to facilitate gripping. Examples include chemical and thermal adhesion. Astrictive methods are based on binding forces between surfaces. Magnetic and electrostatic adhesion and vacuum suction can lift most objects even without direct initial contact. Adaptive grippers possess either integrated or external monitoring sensors which imply the need for specific data processing techniques. Table 1, repeated here from [Iberall, 1997] for simplicity to the reader, provides a good representation of gripping methods and their non-penetrating and penetrating designs.

Table 1 – Gripping Method with non-penetrating and penetrating designs [Iberall, 1997]

Gripping Method	Non-penetrating	Penetrating
Impactive	Clamping jaws, chucks, collets	Pincers, pinch mechanisms
Ingressive	Brush elements, hooks, hook and loop (Velcro)	Needles, pins, hackles
Contigutive	Chemical adhesion (glues), surface tension forces	Thermal adhesion
Astrictive	Electrostatic adhesion	Magnetic grippers, vacuum suction

c. Survey Objective

The objective of this survey is to educate the manufacturing community about industrial manipulation capabilities, advancements, and research focused on autonomous assembly applications. The paper addresses advanced dexterous manipulation by surveying the state-of-the-art in end-effector and manipulation capabilities for autonomous assembly tasks.

d. NIST Mobility and Manipulation Project

The National Institute of Standards and Technology's (NIST) has been researching measurement science and standards for automated manufacturing since the 1980's, beginning with the Automated Manufacturing Research Facility (AMRF). [Zenzen, 2001] Recently, NIST began the Measurement Science for Intelligent Manufacturing Robotics and Automation (MSIMRA) Program which seeks new measurement science to gauge the performance of intelligent automation systems in a variety of areas, from perceiving movement in their vicinity to meeting production goals to protecting nearby humans. Such flexible systems will improve safety and make it easier and quicker to build new products. Within this NIST program is the NIST Mobility and Manipulation Performance Measurements and Standards (MMPMS) Project [MMPMS, 2011] which is tasked with researching measurement science for advanced, dexterous robot manipulation for collaborative human/robot assembly operations. An example of human/robot collaboration is a robot grasping a heavy part for fastening to an assembly, while a human performs the more dexterous operation of bolting the part into place.

Advancing industrial manipulation systems towards autonomous assembly applications has been requested by robot manufacturers to improve the capabilities of these systems and by automobile manufacturers to apply these improved systems. Initially, the project must understand the state of the science and uncover gaps in research, technology implementation, and standards.

The MMPMS project has designed, developed, and implemented a flexible robot testbed at NIST as shown in Figure 2. The testbed includes an industrial robot, currently configured underslung, though the robot could also be mounted upright or on a pedestal. A NIST-developed automated guided vehicle (AGV) is also included in the testbed, along with conveyers, mannequins, test equipment, sensors, and computer systems. A variety of measurement systems are used in the testbed; for example, high accuracy laser tracker with metrology targets, ceiling-mounted barcodes and stereo camera, and wall-mounted reflectors with spinning laser positioning. The testbed is currently being used, among other areas, to research:
- automated guided vehicle safety to improve the American National Standards Institute/Industrial Truck Standards Development Foundation (ANSI/ITSDF) B56.5 safety standard [ANSI, 2011],
- robot safety to improve the International Organization for Standardization (ISO) 10218 robot safety standards and develop the ISO TS 15066 technical specification of collaborative robots [ISO, 2011],
- virtual and real pallet assembly to develop standardized test methods for robotic palletizing
- perception control to develop the ASTM WK 31638 - pose measurement systems standard [ASTM, 2011]

Next, and based on the findings of this survey, the testbed will be used for developing new measurement science for parts assembly tasks and improving associated standards.

Figure 2 – NIST Measurement Science for Intelligent Manufacturing Robotics and Automation (MSIMRA) Program Testbed.

2. Current Robotic Assembly Systems

Combinations of sensing and other devices add value to the robot and manufacturing assembly process, but also complicate the robotic system. Figure 3 shows a block diagram of several sensors and devices that could be added to a robot to provide additional safety and functionality, including: tool changing, compliance, force and torque measurement, structured light, and other sensors.

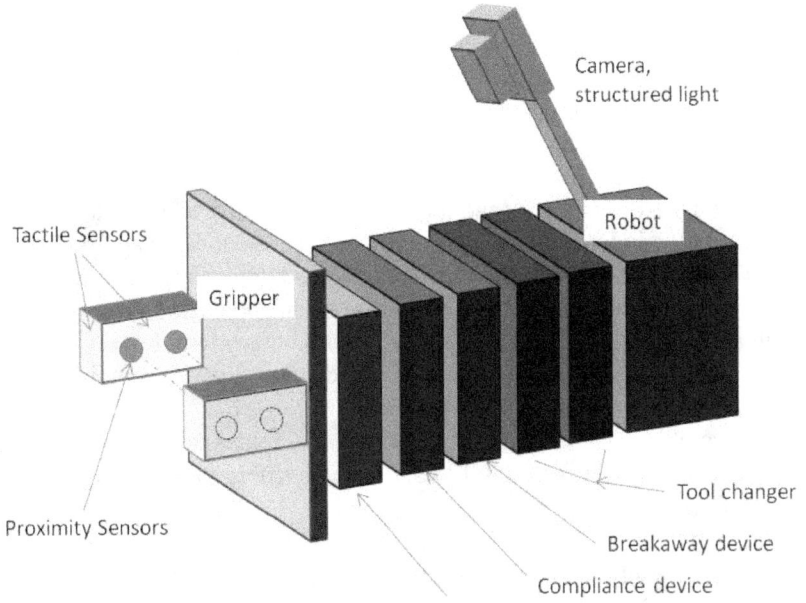

Figure 3 – Block diagram of a robot end-effector showing several sensors and devices that could be attached to a robot to provide additional functionality.

Tool changers are used by robot users to allow a single robot to accomplish more tasks by automatically changing end-effectors or other peripheral tooling. Recent advancements include internal channeling to allow cabling and pneumatics to pass through the unit adapting to new hollow robot wrists. Robotic collision sensors (breakaway devices) prevent damage to robotic end-effectors resulting from robot crashes. Their features can include: automatic reset, high repeatability, large moment rotation, rugged design, and low cost. Compliance devices allow a robot to compensate for positioning errors due to machine inaccuracy, vibration, or tolerance, thereby lowering contact forces and avoiding part and tool damage. Examples of these three devices are shown in Figure 4.

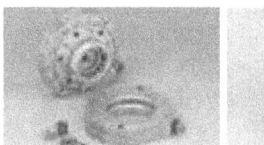  

Figure 4 – Robot tool changer (left), collision sensor (middle), and compliance device (right) (www. ati-ia.com)
[photo-use permission granted by ATI-IA]

a. Gripper Types

i. Class I – End-effectors with fixed shape fingers

Class I end-effectors used for assembly, namely grippers, shown in Table 2, can manipulate and hold a variety of parts, including: round, square, and flexible (hoses, cables – conformable parts). The gripper is designed to ensure that the part it holds cannot translate or rotate with respect to the gripper (i.e., constrained in all six degrees of freedom). Grippers typically use friction, physical constraint, attraction, or support to hold parts. Class I grippers are primarily friction or physical constraint types with soft, replaceable material between the gripper jaw and the part.

Class I special purpose grippers, such as grippers with replaceable jaws, are specific to the part being carried and manipulated. A variety of special purpose grippers, as exemplified in Table 2, are designed to accurately pick up single or multiple parts.

Table 2 – Class 1 Gripper Examples

Gripper Class Example	Brief Description
	Two fingers - Revolute (angular) and translational pairs: (left) Rotational or angular and (right) translational (www.schunk.com) two fingered grippers. The translational gripper shows the universal adapter (bottom tabs) ready for attaching custom fingers as in the extended jaw case that follows. *[photo-use permission granted by Schunk]*

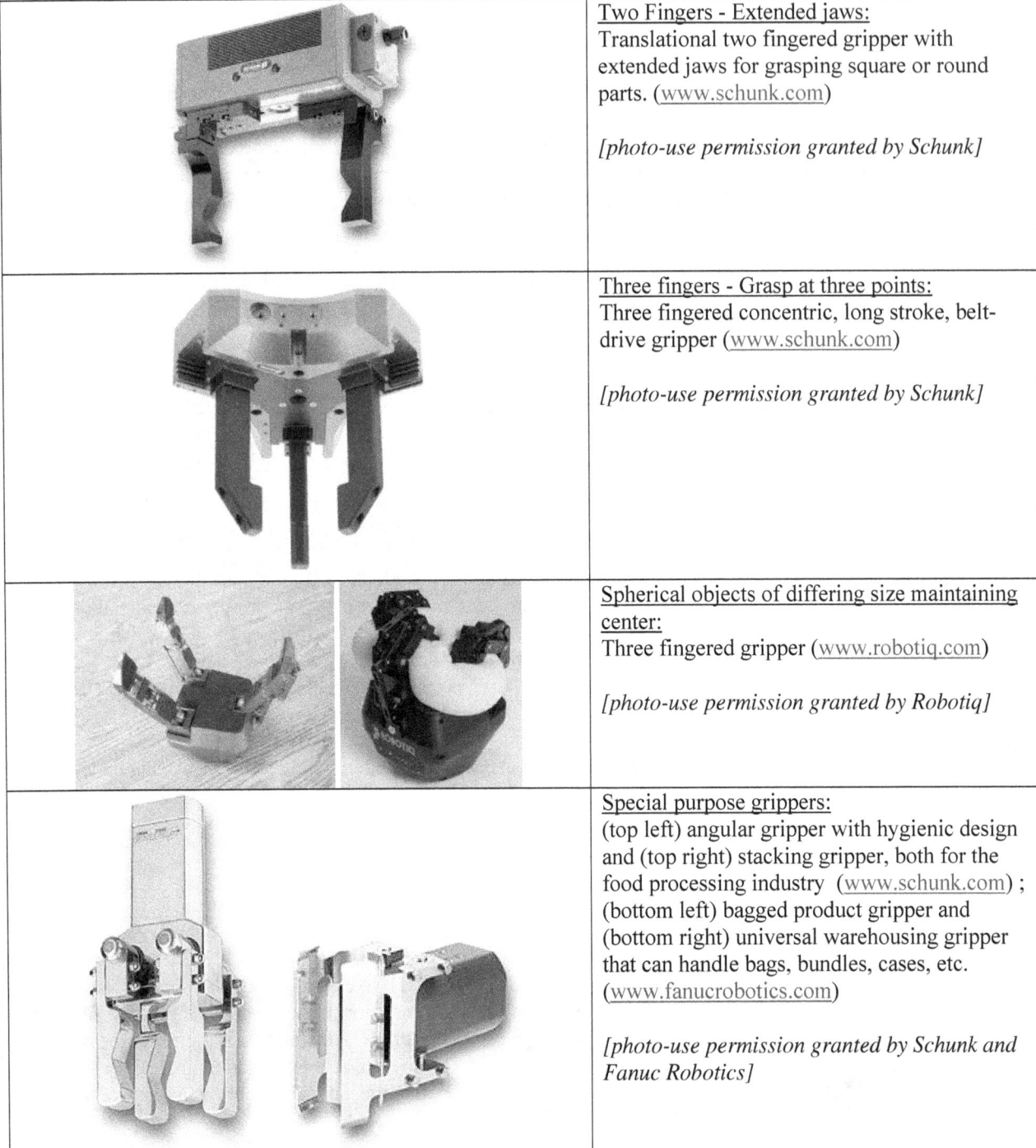

	Two Fingers - Extended jaws: Translational two fingered gripper with extended jaws for grasping square or round parts. (www.schunk.com) *[photo-use permission granted by Schunk]*
	Three fingers - Grasp at three points: Three fingered concentric, long stroke, belt-drive gripper (www.schunk.com) *[photo-use permission granted by Schunk]*
	Spherical objects of differing size maintaining center: Three fingered gripper (www.robotiq.com) *[photo-use permission granted by Robotiq]*
	Special purpose grippers: (top left) angular gripper with hygienic design and (top right) stacking gripper, both for the food processing industry (www.schunk.com) ; (bottom left) bagged product gripper and (bottom right) universal warehousing gripper that can handle bags, bundles, cases, etc. (www.fanucrobotics.com) *[photo-use permission granted by Schunk and Fanuc Robotics]*

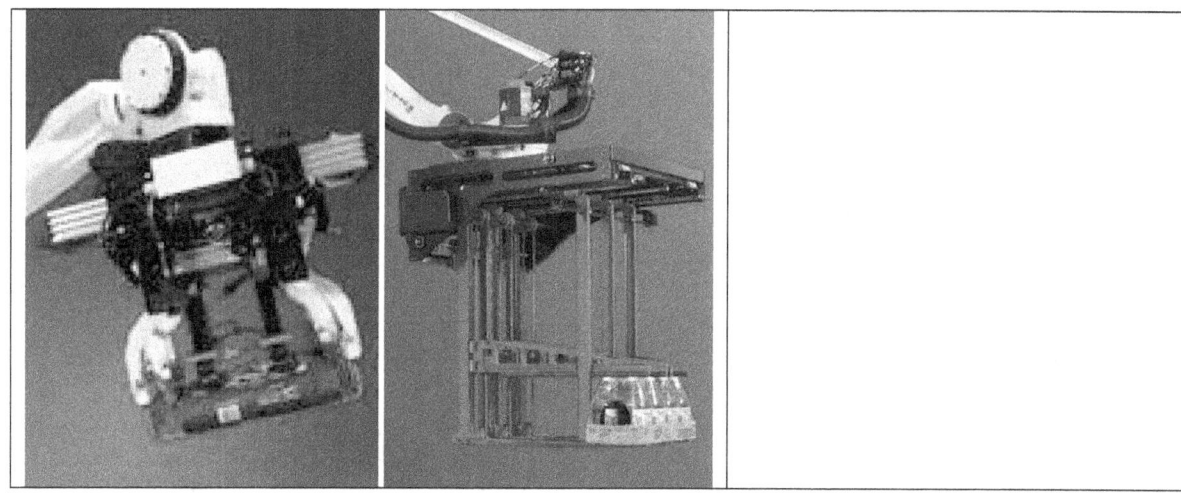

ii. Class II - Special purpose end-effectors

Table 3 shows examples of Class 2 special purpose end-effectors and provides a brief description of each. There is a huge variety of suction cup style vacuum grippers. Shapes include oval and round and sizes can range from 12.7 mm (0.5 in) up to 101.6 mm (4 in) in diameter. They can be made of Viton fluoroelastomer, silicone, other rubbers, or polyurethane. Picking the right cup material, size, and shape requires experience, vendor guidance, and sometimes trial and error. Even with all of the variety available in cups, they won't work in every application. Sometimes holes in the part preclude use of vacuum cups. Cups also can cause cosmetic problems on Class A surfaces - freeform surfaces of high efficiency and quality. Rough surfaces also may not work well with cups. In any of these cases, alternative ways to secure parts are needed.

Vacuum cups typically engage in one of three ways: at the front or side of the part or at an angle. For side engagements, short-stroke pneumatic cylinders advance the cups to grip the part after the vacuum cup frame moves into position. For quick pickup and placement of parts or for fast cycles, spring arms allow the cups to be in place before ejection and then accept the part like a mitt accepts a baseball. Hex-shaped gripper-arm profiles will keep grippers from turning when their orientation to the part must remain fixed. [EOAT, 1997]

Electro and permanent magnet devices are used to pick up and release ferromagnetic parts. Passive magnets are typically used as an element with other class II gripper technologies. Electromagnet type grippers gain an attractive force only when current is applied to the unit.

Radiant energy joining processes include electron beam and laser welding which focus an energy beam on the work pieces to be joined. We focus on laser welding or joining as it is more advanced and has many advantages over electron beam welding. Below is an example list of advantages of laser beam welding where a complete list is shown in [Sivam, 2004]:
- Welds can be made inside transparent glass or plastic housings.
- A wide variety of materials can be welded, including some formerly considered as unweldable combinations.
- As no electrode is used, electrode contamination or high electric current effects are eliminated.
- Areas not readily accessible can also be welded.
- It permits welding of small, closely spaced components with welds as small as a few microns in diameter.
- Unlike electron beam welding, it operates in air and no vacuum is required.
- Since the laser beam is highly concentrated and narrowly defined, it produces a narrow size for the heat affected zone.
- Because it is light, it is clean and no vaporized metal or electrodes dirty up the delicate assemblies.

Spray adhesive has many temporary and permanent assembly applications including, for example: application of die lube, mold release, and glue for metal and plastic parts, spraying release agent on screens and drums in wood building panel manufacturing, applying adhesive to packaging materials to prevent slippage on pallets, applying adhesive on tire treads.

Robotic deburring tools are robotic end-effectors for removing burrs, flashing, and other unwanted edge properties caused by cutting or machining and are often used as an integral part of the assembly process. There are several basic types of deburring end-effectors. Radially-compliant robotic deburring tools have a rigid outer housing and internal motor/spindle assembly that mounts on a pivot bearing. Pistons in a chamber near the front of the housing supply a constant flow of air pressure to a rotating shaft. A rotary cutting burr or file rides on a cushion of air that provides a reliable field of compliance while maintaining a constant force and spinning at high speeds. This field of compliance is exerted in the radial direction, providing a high degree of stiffness in the path direction and a low degree of stiffness in the contact direction. Often, a floating head is used to compensate for variances in robot path or part position. Axially-compliant robotic deburring tools are also available. These devices exert a constant axial force on a deburring head mounted to a free flying piston (FFP). The movement of these high-torque tools compensates for changes in part tolerances, part misalignment, and robot path variation.

Table 3 – Class 2 Gripper Examples

Gripper Class Example	Brief Description
	Vacuum gripper: (top) supporting and placing a box on a conveyer at NIST. This gripper was custom designed from a series of pneumatic components and cups similar to those shown (bottom) (www.anver.com) *[photo-use permission granted by Anver]*

	Electromagnetic gripper: Magnetic gripper handling piston rods (Kawasaki Robotics - http://www.youtube.com/watch?v=eGPne8_sR4c)
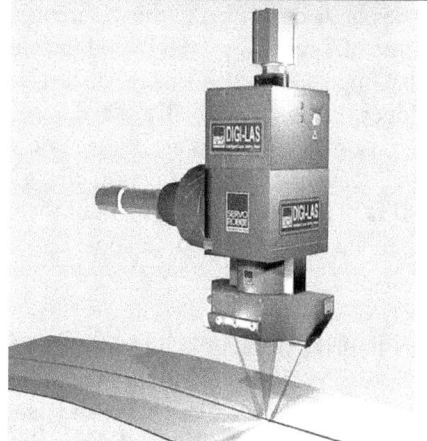	Welding heads: Intelligent laser welding head (www.servo-robot.com) *[photo-use permission granted by ServoRobot]*
	Machining Spindles: A product line of robotic deburring end effectors. (http://www.ati-ia.com/products/deburr/deburring_home.aspx) *[photo-use permission granted by ATI-IA]*

	**Sprayers:** Robot spraying adhesive to mount headlights in automobiles during manufacturing. (www.kuka.com) *[photo-use permission granted by Kuka]*

iii. Class III - Multipurpose end-effectors

Table 4 shows examples of Class 3 special purpose end-effectors and provides a brief description of each. Advanced (or sometimes called 'universal') grippers offer flexibility to pick up a wide range of objects. Typical advanced grippers are at least three fingered and multi-jointed and are similar in shape and functionality to human hands. This design adds complexity, yet promotes flexibility towards prehension of parts with varying properties.

[Mindtrans, 2010] provides a compilation of figures and descriptions of 22 anthropomorphic robot hands/arms, including prosthetic hands/arms that range in price from $6000 USD to more than $100 000 USD. Three of the 22 described in [Mindtrans, 2010] are exemplified here. One of the hands described in this reference is the DLR-HIT Hand II shown in Table 4 and developed by the Harbin Institute of Technology (HIT) and the German Aerospace Center (DLR). The DLR-HIT Hand II, has five modular fingers and each finger has four joints and three degrees of freedom. Altogether there are 15 motors in the finger body and palm. The hand is actuated by commercial flat brushless DC motors commutated by digital hall sensors. There is an absolute angle sensor and a strain-gauge based joint torque sensor at each joint. The high-speed, real-time communication bus is implemented by field programmable gate array (FPGA). The DLR-HIT Hand II is the further development of the DLR-HIT Hand I, also shown in [Mindtrans, 2010], which they claim has been successfully used in some research institutes in USA, Spain, Italy, Greece, Germany, and China. [DLR, 2011]

The Delft hand/arm is shown in Table 4 and is quite different from the HIT or other anthropomorphic hands and arms from [Mindtrans, 2010]. This system includes a potentially low-power and safe manipulator similar to a classic desk lamp. In contrast with conventional factory robots, the Delft arm has a low mass and uses low-power motors to reduce its cost and be low in mass for additional safety. The design is based on the principle of static balancing, whereby adjustable springs compensate for the mass of the arm. Claimed specifications include:

- masses are fully balanced using adjustable springs
- 4 degrees of freedom
- 4 low-power motors
- 4 rotary encoders
- 10 kg total mass
- 2 kg maximum payload

The Delft hand, suggested in the reference, is designed to be a powerful, versatile, lightweight three-fingered, under-actuated gripper. The three fingers of the hands each have two degrees of freedom, all actuated by a single motor. Special mechanisms assure a powerful and robust grip on widely varying objects with a minimum of sensing and control. Claimed hand specifications include:

- 3 x 2 degrees of freedom
- 1 low-power motor
- 1 force sensor
- 0.6 kg total mass
- 8 kg maximum payload

The Robotics and Mechanisms Laboratory (AKA the RoMeLa Project) at the College of Engineering at Virginia Tech University has designed and built a prototype robotic hand that is controlled and operated by compressed air. Called RAPHaEL (Robotic Air Powered Hand with Elastic Ligaments), the robot can hold heavier, solid objects, as well as light or delicate ones such as a light bulb or an egg. The hand is powered by a compressor air tank at 414 kpa (60 psi) and an accordion-style tube actuator, with microcontroller commands operating and coordinating the movements of its fingers. It uses no other motors, and the strength of the grasp is controlled by a change in air pressure, making the hand quite dexterous.

Table 4 also shows two advanced grippers that conform to the part they are grabbing, rather than being designed for particular parts. The "ball gripper," a project supported by the Defense Advanced Research Projects Agency (DARPA) and including researchers from Cornell University, the University of Chicago, and iRobot Corporation, uses everyday ground coffee and a latex party balloon, bypassing traditional designs based on the human hand and fingers. An everyday party balloon filled with ground coffee -- any variety will do -- is attached to a robotic arm. The coffee-filled balloon presses down and deforms around the desired object, and then a vacuum sucks the air out of the balloon, solidifying its grip. When the vacuum is released, the balloon becomes soft again, and the gripper lets go. [Cornell, 2010]

[Prahlad, 2011] presented a new approach, not shown, to gripping called electro-adhesion which uses electrostatic cohesion to lock onto objects and conformance to encompass an object to grasp and then lock the joints in order to trap the object. This approach works best when large forces are not necessary to keep the object confined.

Table 4 – Class 3 Gripper Examples

Gripper Class Example	Brief Description
	Robot hand: DLR-HIT 5 finger hand *[photo-use permission granted by DLR-Hit-Hand, Robotics and Mechatronics Center]*
	Robot hand: Delft hand/arm (top) and hand (bottom) *[photo-use permission granted by TU Delft; photo by David Joosten]*

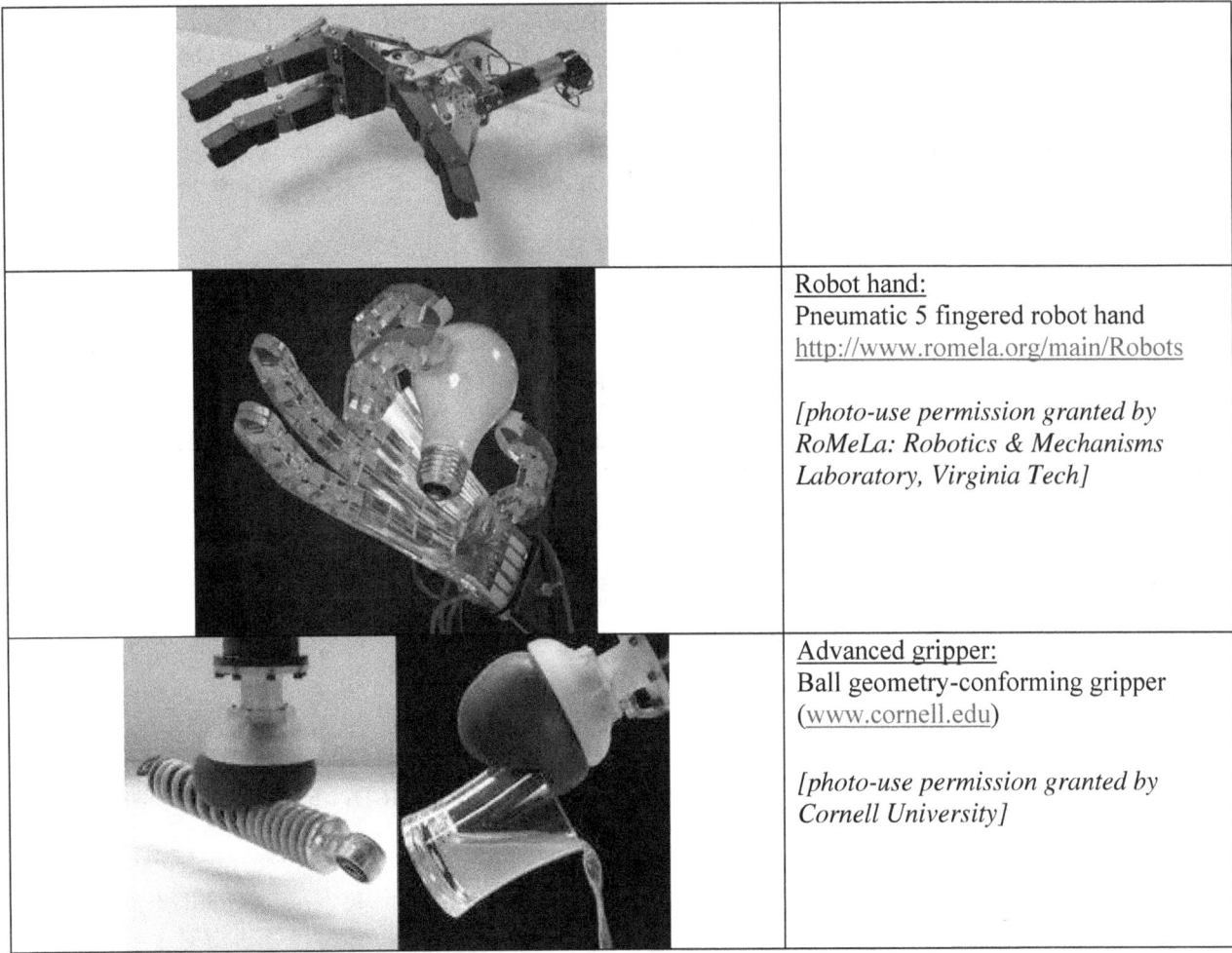

		Robot hand: Pneumatic 5 fingered robot hand http://www.romela.org/main/Robots *[photo-use permission granted by RoMeLa: Robotics & Mechanisms Laboratory, Virginia Tech]*
		Advanced gripper: Ball geometry-conforming gripper (www.cornell.edu) *[photo-use permission granted by Cornell University]*

b. End-Effector Sensing

[Brumson, 2011] says that robots can now perform tasks better because of advancements in sensors. "Applications are moving toward tactile force feedback and non-tactile sensing for positional and robotic guidance. Advancements in vision sensors make them less expensive and more powerful," says Nicholas Hunt, Product Support Manager at ABB Inc. (Auburn Hills, Michigan) "Advancements include high accuracy of vision sensors such as laser scanners. Sensors are very affordable and robot manufacturers are able to incorporate those features and functions inside the robot to accommodate data."

Robots are no longer simply grabbing parts and forcefully moving them into fixtures, asserts Hunt. "With tactile feedback available and the processing power within robots, integrators now allow the process to control the robot's behavior, not the robot's behavior controlling the process. Tactile feedback and vision systems with scanning lasers enable end-effectors to move on a micron level of accuracy."

i. Tactile

Much research has been performed in the area of tactile sensing for robots and is nearing the capability for use in assembly. [Howe, 1993] provides references for robot hand tactile sensing and states that "transmission dynamics such as friction, backlash, compliance, and inertia make it difficult to accurately sense and control endpoint positions and forces based on actuator signals alone." He also lists the common robot hand sensors as: dynamic tactile sensor, tactile array sensor, finger tip force-torque sensor, and joint angle sensors. [Cutkosky, 1993] et. al. describes tactile and force and torque event driven dexterous manipulation with the events primarily determined through tactile and force/torque sensing. They breakdown a simple task (e.g., grasping a glass of water, lifting it and replacing it) that contains several events and discontinuities. Experiments with human subjects reveal that during such tasks people rely on a combination of fast- and slow-acting tactile sensors to detect such events as contact, the onset of motion, and the onset of slipping. Preliminary experiments with a simplified robotic hand have suggested that a combination of force sensors and dynamic tactile sensors can provide robots with a similar ability.

Although there has been much tactile sensor research, we found that relatively few tactile (including spatial measurement and touch) sensors are being used in robotic assembly processes today, perhaps with the exception of sensors to track weld seams. [Kuka, 2011] With any deviations in the weld seam from the model, the sensor tracking the actual seam informs the robot to make error corrections in real time. Currently no general specification of a touch or tactile sensor exists. Crowder suggests that the following can be used as a basis for defining the desirable characteristics of a touch or tactile sensor suitable for the majority of industrial applications [Crowder, 1998]:

- A touch sensor should ideally be a single-point contact, although the sensory area can be any size. In practice, an area of 1 mm^2 to 2 mm^2 is considered a satisfactory compromise between the difficulty of fabricating a sub-miniature sensing element and the coarseness of a large sensing element.
- The sensitivity of the touch sensor is dependent on a number of variables determined by the sensor's basic physical characteristic. In addition the sensitivity may also be the application, in particular any physical barrier between the sensor and the object. Sensitivity within the range 0.4 N to 10 N, together with an allowance for accidental mechanical overload, is considered satisfactory for most industrial applications.
- A minimum sensor bandwidth of 100 Hz.
- The sensor's characteristics must be stable and repeatable with low hysteresis. A linear response is not absolutely necessary, as information processing techniques can be used to compensate for any moderate non-linearities.
- As the touch sensor will be used in an industrial application, it will need to be robust and protected from environmental damage.
- If a tactile array is being considered, the majority of applications can be undertaken by an array of 10 to 20 sensors square with a spatial resolution of 1 mm to 2 mm.

Commercial tactile sensors that are only recently being implemented are from [Pressure Profile, 2010]. They have developed the FingerTPS - Tactile Pressure Sensing for the Human Hand (see Figure 5 left) which uses capacitive-based pressure sensors to quantify forces applied by the human hand through finger gloves and a wireless data transfer system. Also for robot grippers, they have developed the RoboTouch Systems Pressure Profile System (see Figure 5 – right) which provides tactile feedback for robot grippers. The technology and company were born out of the Robotics Laboratory at Harvard University over a decade ago. Currently, three gripper and robot companies have implemented this tactile system. RoboTouch can include multiple sensing pads, each with 12 to 24 sensing elements with digital interfaces. These can be placed on the robot fingertips, grippers, and/or palm surfaces.

Figure 5 – Tactile sensors for the hand (left) or robot figures and/or palm (right) (www.pressureprofile.com)
[photo-use permission granted by Pressure Profile]

Tactile sensors can complement visual systems by becoming the controlling system at the time contact is made between a gripper of the robot and an object or objects being gripped, a point when vision is often obscured. [Wikipedia (tactile sensors), 2010]

ii. Proximity

There are several non-contact sensing devices (proximity devices) typically used with robots to measure and provide the distance to objects to pick up or avoid. The types of sensors are [Wikipedia (proximity sensors), 2011]:

- Inductive
- Capacitive
- Capacitive displacement sensor
- Eddy-current
- Magnetic, including Magnetic proximity fuse
- Photocell (reflective)
- Light or Laser rangefinder
- Sonar (active or passive)
- Radar
- Passive thermal infrared
- Passive optical (such as charge-coupled devices)
- Reflection of ionizing radiation

Although not for assembly tasks, [Volpe, 1994] surveyed proximity sensors for use in manipulator collision avoidance. Five categories of sensors were considered for this use in space operations: intensity of reflection, triangulation, time-of-flight, capacitive, and inductive. From these categories, the most promising commercial and mature laboratory prototype sensors were triangulation, time-of-flight, and capacitance sensors.

NIST developed a proximity sensing system embedded in robot gripper fingers to provide part presence and centering as well as gripper rotation information to the robot. Figure 6 shows pairs of infra-red proximity emitters and detectors embedded into machined plastic fingers. Four pairs of emitters and detectors were embedded in the finger pads. The sensors provide part presence and centering capability that can update robot position with respect to the part. Also, front and bottom pairs of emitters and detectors provide tool rotation with respect to parts trays. Left and right emitter and detector pairs sensing similarly reflective surfaces from the front sensors (fingers pointing down) and bottom sensors (fingers parallel to a tray) are input to the robot controller, compared to each other, and used to level the gripper with respect to the parts tray prior to picking up parts. A similar system used both infrared and ultrasonic sensors embedded in the robot gripper fingers. [Bostelman, 1989]

Figure 6 – Gripper fingers with pairs of infra-red proximity emitter and detectors embedded into machined plastic fingers (NIST).

To date, most machine control applications have used inductive proximity switches. Based on the electromagnetic induction principle, these sensors are designed to detect metal targets and are generally insensitive to the effects of dirt. A major drawback, however, is the cost and quantity of cabling needed to connect each sensor to a factory's control system. Commonly located on moving parts of machines, these types of sensors are prone to malfunction due to cable damage caused by wear and tear. ABB developed a wireless proximity switch, the WISA (Wireless Interface for Sensors and Actuators) protocol, and a power system that eliminates the need for cabling in proximity sensor applications. [ABB, 2005] Data is relayed from the switch to the machine control system using ABB's WISA protocol, specifically designed for industrial applications.

Most proximity sensors used in manufacturing are not attached to the robot, but rather to conveyers, support structures, or other areas. They are typically used to detect the presence of parts. The latest is high speed, colored parts detection as shown in Figure 7. Unlike vision sensors that are designed for pattern detection, contour verification, or edge location, full color photoelectric sensors are aimed only at a specific spot on the target that verifies that the right product or the desired attribute is present. This lets them operate at speeds as fast as 1 ms, which is much faster than the typical 20 ms update time required by vision sensors. [Draper, 2008]

Figure 7 – Color photoelectric sensors detecting parts on a conveyer.
[photo-use permission granted by Balluff]

iii. Force/Torque

Force/torque sensors are used in adaptive control schemes for part assembly operations, constant force operations such as buffing, polishing, and deburring. They are also used in a passive mode to collect force data for lot testing and statistical process control (SPC). [Perry, 2002] Force/torque sensors can be in-line with the robot end-effector as shown in Figure 3, carried by the robot as in servo guns for welding operations, or measured at each joint by independent torque sensors [Motoman, 2006] or through AC drives on robot joints [ABB, 2011]. Examples of some robot force controlled applications follow. Appendix D provides a list and drawings of current force control function types and descriptions. [Fanuc, 2007]

Kuka Robotics Corporation has delivered systems that perform grinding and milling operations with the use of force and torque sensors. [Ogando, 2007] Force controlled robots are starting to become more popular in "premachining" applications--or the use of robots to perform rough machining operations, leaving only a single pass on a computer numerically controlled (CNC) machine tool for finish machining. In this case, force control helps the robot close the gap with machine tool feeds and speeds by optimizing the contact forces between the robot-borne tool and the workpiece. Figure 8 (left) shows a robot arm fitted with a force/torque sensor and checks the actuation forces on an automobile cruise-control switch.

ABB Robotics has introduced a new system of robot control for assembly applications. The ABB RobotWare Assembly Force Controller utilizes force/torque sensors, adds sensor feedback to the robot's positioning, and allows the robot to search for the correct assembly position. Forces and torques are measured by the sensor at the wrist of the robot giving it a tactile sense of touch. This system makes it possible to automate tasks that earlier required skilled personnel. Figure 8 (right) shows a robot performing assembly of automobile engine cylinders. Fanuc Robotics has a similar product with force sensing to provide 3D assembly with six degrees-of-freedom (6DOF).

In another application, ABB Robotics utilized a force and torque sensor to unwind, slit, and then re-wind paper to new dimensions. The robot, equipped with the sanding head, smooths the edges utilizing the force/torque sensor technology to provide force feedback. This enables the robot to feel and have a sense of touch, just as a human would. This sense of touch allows the robot to make quick adjustments in real-time to maintain a constant contact force. [ATI, 2007]

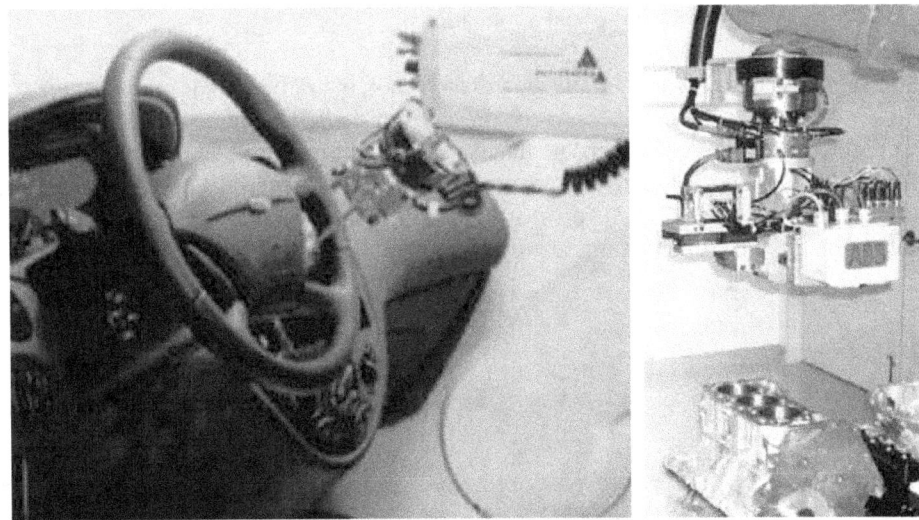

Figure 8 – (left) A robot arm fitted with a force/torque sensor checks actuation forces on an automobile cruise-control switch. (MachineDesign.com), (right) a robot performing assembly of cylinders (www.abb.com)
[copyright photo-use permission granted by ATI-IA and ABB]

iv. Vision

For current machine vision to work effectively, engineers must make sure the parts have a consistent visual appearance. They can also include features that enable easy recognition. For example, engineers can add a boss or other feature to a double-sided part to help a vision system distinguish one side from another. [Assembly, 2005] However, [Schofield, 2010] suggests that "the use of vision guided and tactile sensing is enabling robots to become more reactive to their environments and less reliant on component characteristics. He suggests that product manufacturing flexibility can be increased with ABB Robotics' automated 'TrueView' vision system with application to engine assembly. It uses a single camera for 3D vision guidance of robots where they suggest benefits to be the elimination of expensive fixtures and it automates operations that previously required human intervention. Claims include that it can improve efficiency, that the robot can react to changes in the environment, that there is no pre-arranging or pre-placing required, and that an auto-calibration can be used for easy system integration.

Fanuc [Fanuc, 2011] sells their SYSTEM R-30iA™ robot controller which comes standard with the Fanuc iRVision hardware. The laser vision system allows the robot to 'see' the location of a part and the system is integrated directly with the Fanuc R-J3iC controller. By loading the vision software option and connecting a camera directly to the main CPU board, the user can add a vision process to the robotic application. See Figure 9.

One example of perception use is from [Kuka, 2011], where different components are assembled on a function carrier in door linings. Car doors now serve many purposes, such as housing window lifters and audio components and measuring the lateral acceleration in the event of an impact for airbag deployment. Robots carry out all handling and assembly tasks for many different function carriers including removing the function carriers from an injection molding machine. An adaptive gripper system enables the removal of all the function carrier variants without the need for a change of gripper. Two cameras secured to the gripper establish the external contours and check the dimensional accuracy of the openings.

In another example [Packworld, 2009], a robot was used to glue a packet or sachet of laundry detergent within a very tight tolerance to a folded card on a promotional mailing piece. The requirement was to pick a sachet in a random location from a moving conveyor and place it onto the card, located on another moving conveyor in a specific location, with ±3 mm (1/8 in) accuracy, at 50 pieces/min. The system used a two-camera, proprietary vision system to pick pieces from random locations and orient them before placing them on a moving target. The camera locates the packets and cards on the conveyors and sends the offset information to the robot controller. The controller then combines this data with the conveyor speed information to track each packet. It then picks the packet with a vacuum cup when it comes within range of the robot arm and places it on the out-feed conveyor.

Figure 9 – Vision system being used to guide the robot to load a box with parts (www.fanucrobotics.com).
[copyright photo-use permission granted by Fanuc Robots]

"Of course, even with today's user-friendly vision technology, setting up a new system can still be challenging. For example, lighting in machine vision applications is still more art than science. Lighting trials done in laboratories rarely duplicate production environments. Moving parts, factory lighting, air quality, outside windows, and skylights can adversely affect the performance of many vision applications. Engineers must ensure that once the lighting has been resolved, factory conditions will not affect a new system after installation."

Varying lighting conditions throughout the day and year, correcting for parallax, and correlating the vision data with the robot controller were other issues to resolve. A constant light source, high camera mount, and camera calibration, respectively solved these issues. [Assembly, 2005]

"Today, the 3D sensor is about where machine vision was 20 years ago. You can buy a 2D vision system that's built into the camera with Ethernet and almost anyone can program the thing, and that's where we need to get with 3D vision." [Hardin, 2005] 3D machine vision's biggest problem is that it's not as easy to use as 2D because it needs a combination of light sources and calibration built into the sensor more than 2D cameras. LMI is tackling the simplicity challenge by offering lines of 3D "smart gauges" that include sensor(s), lighting, processing, and calibration in one package. Some smart gauges have more than one camera – stereoscopy with structured light. Others use lasers for fast applications because of the light intensity that lasers provide. LMI's newest sensor does full 3D imaging in full color. Each sensor has multiple cameras and multiple light sources at different frequencies and the system combines the image automatically to provide a mixture of 3D and color. Lasers are suggested as the most robust method since they are not influenced by ambient light. In addition to lighting and computing developments, improved auto-calibration routines in 3D machine vision are an enabling development for 3D machine vision that is expanding its utility. SICK Inc. says their IVC-3D system helps to simplify 3D measurements by placing image filtering circuitry on the same chip with a CMOS optical sensor. The IVC-3D also uses laser line triangulation to create 3D data sets, but the processing on the CMOS chip allows the sensor head to isolate the laser line within the 512 pixel x 1536 pixel image and only transmit the 3D information with sub-pixel accuracy. This allows the system to run at extremely high frame rates up to 5 000 frames per second (fps) at full frame. Fast 3D analysis is critical to new vision applications that track moving objects, such as furniture on moving chain hooks, for example, or mobile robots on carts. Geometry based recognition systems can recognize more variations, e.g., changes in lighting, focus, feature, shape, etc. than laser based systems. They are key to handling bin picking, auto-tracking, and products that can change shape.

c. Generic Robot System Assembly Tasks

Autonomous assembly typically requires coordinated robot manipulation combined with end-effectors (e.g., gripper) and sensing (e.g., force, vision, tactile). A more detailed view of research on several generic robot tasks follows, including insertion and bin-picking. Robot bin picking uses vision and a robot to locate and pick parts from a bin or moving conveyor and eliminates the need for collating, accumulating, and orienting. The more generic 'material handling' is also used in the assembly process, although it is not discussed in this paper beyond the information presented in Section 2a Gripper Types that allow material handling by robots.

i. Insertion

Peg-in-hole insertion or fitting is not only a longstanding problem in robotics, but the most common automated mechanical assembly task. [Paulos, 1993] Both force and vision are being used to support peg-in-hole insertion. Passive force assembly is supported by chamfers on mating parts or on remote center of compliance devices. The purpose of force control is to make the robot sensitive to contact forces. A basic endeavor in robot force control is how to determine the interaction forces and efficiently use the feedback signals in order to synthesize the appropriate input signals, so that the desired motion and force can be maintained. The basic variables in robot force control are position, velocity, acceleration, and force. The differences in the existing fundamental force control algorithms stem from the different application of these basic variables and their relationships. The force control result is that the robot can "feel" its surroundings. It can apply a constant force on a surface, even if the exact position of the surface is not known. [ABB, 2006] Several software components are typically included in robot options with force control, including: gravity compensation, sensor offset calibration, activation and deactivation of force control, reference values definitions (desired force, torque or movement), end conditions, supervision, force/load monitoring, and data types. Several robot manufacturers offer optional force control. An example of force control function types is shown in Appendix d. A basic approach to force control is as follows:

1. Identify the load and calibrate the system.
2. Set up desired force and movement pattern.
3. Set up end condition.
4. Activate force control.
5. Activate force and movement pattern.
6. Wait for end condition to occur.
7. Deactivate force and movement patterns.
8. Deactivate force control.

Paulos explains a method for high precision, self-calibrating, peg-in-hole insertion using several very simple, inexpensive, and accurate optical sensors. (see Figure 10 top). The concept was tested on a static part and dynamic peg being inserted into the part. The sensors were simple optical beam sensors, which responded to the presence or absence of an object along the beam line. The self-calibrating feature allowed successful dead-reckoning insertions with tolerances of 25 micrometers without any accurate initial position information for the robot, pegs, or holes. The implemented program worked for any cylindrical pegs, and the sensing steps do not depend on the peg diameter, which the program does not know. The key to the strategy is the use of a fixed sensor to localize both a mobile sensor and the peg, while the mobile sensor localizes the hole. The strategy is said to be extremely fast, localizing pegs as they are en route to their insertion location without pausing. The result is that insertion times are dominated by the transport time between pick and place operations.

More recently, [Youngrock, 2008] performed dynamic peg and dynamic part (i.e., moving peg and part) peg-in-hole insertion using vision as shown in Figure 10 bottom. Their real-time visual servoing system includes tracking the part with a camera and also a stereo camera shown in Figure 8 (bottom, left). The stereo camera is used by another visual tracking module that is a part of a multiple-vision-loop architecture. Purdue University demonstrated a Kalman-filter-based framework [DeSouza, 2004; Yoon, 2008] that carries out fast and accurate rigid object tracking even when the object motions are large and jerky.

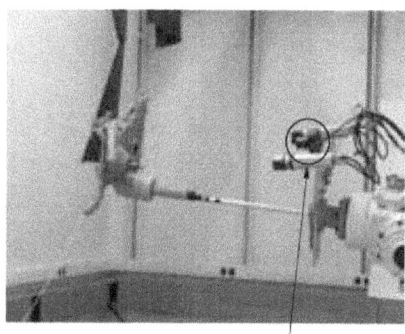
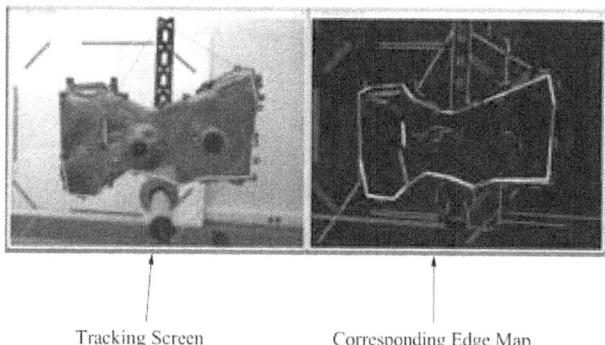

Figure 10 – Robot assembly showing (top) peg-in-hole insertion into a static part (bottom) and in a dynamic swinging part (www.purdue.edu) *[photo-use permission granted by Purdue University]*

A type of peg-in-hole insertion utilizes threaded screws. Approximately a quarter of assembly operations in the manufacture of commercial products are screw insertions. [Lara, 1998] However, we found very little information about robot manufacturers and users programming and implementing robots to thread screws for assembly of parts. There has been some research in this area that will be presented later in Section 3b(iv) on threaded insertion. A single reference from [Masi, 2006] briefly discusses the coordinated action of two robots assembling a gear case. He describes that positioning errors between the grippers can and will lead to stripped screw threads and broken components. For coordination, Fanuc developed a multi-robot controller along with a robot simulator where one controller has access to all sensor information from all of the load cells and encoders on both robots simultaneously. One robot picked up one half of the gear case housing and presents it to the other robot. The second carefully inserts and meshes gears on a bearing shaft already in the housing. The second robot then picks up the housing's second half and locks it in place over the assembly. The two robots then reorient the housing to present it to a third robot which inserts and drives screws that ultimately hold the gearcase together. All three robots used machine vision to accomplish the assembly task.

ii. Bin Picking

In many industrial processes, component to be assembled are delivered scrambled in boxes. Usually these parts must be picked out of the box manually to feed them into an automated process. Using an industrial robot for this task is very difficult. Robot vendors including Fanuc, Motoman, and Staubli have recently demonstrated bin-picking systems at trade shows. At least one North American systems integrator planned in 2006 to offer "semi-random" and random bin-picking as "standard product" technology for certain types of parts. A number of automotive industry end-users are also experimenting with bin-picking technology, with some early applications already in production. [Iversen, 2006] [Shafi, 2007] explains how to implement bin-picking in a manufacturing operation.

[Yaskawa, 2011], developer of Motoman robots, claims to have a cost-effective, modular packaging solution that is ideal for small cartons, bottles, and pouches. Multiple configurations are offered to meet specific throughput requirements. Figure 11 shows Motoman's high-speed, six-axis robots placing products into cases. Systems are available for partitioned cases. Reconfigurable grippers can accommodate a range of consumer products and can make multiple picks at a time.

Figure 11 – Motoman robot placing parts into a box (www.motoman.com)
[photo-use permission granted by Motoman]

[Delden, 2006] discussed an approach to grasping and recognizing parts in an industrial robotic work cell. Figure 12 shows examples of (left) a bin of parts and (right) a robot picking parts from a bin. The centroid, orientation, and length of elongated parts lying on a flat work area are estimated by a sequence of simple algorithms. Off-the-shelf components and freely downloadable software application programming interfaces (API) attempt to make the system inexpensive and easily implemented. The approach has been implemented and tested with an industrial robot. However, retrieving randomly placed parts from a bin is much less structured and more process intensive.

 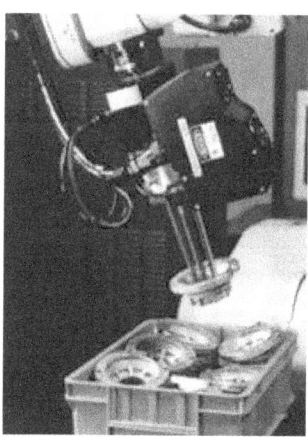

Figure 12 – (left) bin of parts (www.tec-automation.com/.../random-bin-picking/), (right) robot picking parts from a bin (www.adaptivesystems-usa.com/integ_assembly.asp)
[copyright photo-use permission granted by Fanuc Robots]

d. Specific Robot System Application Examples

Many robot assembly tasks are currently being performed, most of which utilize fixtures, clamps, and/or jigs. Few tasks include robots that can adapt to flexible processes. The following is an example list of assembly tasks that robots are currently performing [Kuka, 2011; Motoman, 2011; ABB, 2011]. The examples are shown here for the reader to realize current and available assembly solutions. Many more examples are available on these and other robot and end-effector manufacturer websites.

- screws stoppers into barrels
- drills and installs fasteners in aircraft components
- cuts increasingly complex holes in cars requiring a high degree of repeatability
- picks up a part from a specially designed transfer system, places it in and retrieves it from a mold
- places the required bearings and then the finished plastic part into the press unit
- assembles fuse switch disconnects from three different pieces, places it in a press, and then mounts a spring and screws in it
- handles parts through trim, bend, assembly, and spot welding operations
- assembles LED lamps for semi-trucks, trailers, and off-road equipment
- two arms are used to insert a sub-assembly into an office chair; one arm holds the chair while the other attaches a bracket
- unloads a set of left- and right-hand parts from a four-station positioner, manipulates parts for bolt insertion, places parts into separate dunnage baskets, and adds slip sheets
- holds a part in its gripper, picks up the next part with the other side of the end-effector, and places them
- feeds the cylinders of the unit with the thread inserts, which it had previously taken out of sorting pots with its suction gripper
- picks up eight thread inserts individually in one cycle and then passes them on as a group to the cylinder
- sews leather covers for seat squabs and seat backs for both basic and multi-function seats
- screws brackets onto the front end of a car
- positions parts in a press unit and presses in two types of quick-action fasteners
- removes the parts from an injection molding machine with an adaptive gripper system for a variety of parts
- checks the thickness of material with measuring tongs

- inserts hubs into the die with 0.02 mm tolerance
- inserts clips into flexible cable ducts

Automobile manufacturers, and perhaps others, are requesting specific dexterous manipulation tasks solved by robots that can be reconfigurable for a variety of vehicles or other equipment, parts, and assemblies being manufactured. One manufacturer suggested the following assembly areas for robot manufacturers to consider automating:
- Wiper module assembly installation
- Headlights fastening including cable connection to the light
- Trimline installation
- Gas, brake, and other fluids tanks filling
- Weather strip installation
- Carpet installation
- Seat installation
- Wheel and tire fastening
- Hoses, cables, and wiring harnesses installation (see Figure 13)

Figure 13 – Examples of automobile harnesses and cables and their relatively difficult-to-access locations

Robotic assembly systems offer tremendous promise for flexible assembly automation, but present a variety of complex research issues due to the positioning inaccuracy of the manipulator, dimensional variation of mating parts, and their physical interactions. [Cho, 1987]

3. Advances in Robotic Assembly

a. Systems

The nature of manufacturing robotics is changing to be more collaborative with humans. Rodney Brooks, Heartland Robotics Founder and Massachusetts Institute of Technology (MIT) Professor, says of the future for robotics:

"Today's manufacturing robots are big and stiff, unsafe for people to be around, engineered to be precise and repeatable, not adaptable. Normal workers can't touch them... What if ordinary people could touch robots? What if ordinary people got to interact with them and use them?"

[Motoman, 2010] developed the unique dual-arm SDA-series (slim, dual-arm) robot that combines 'human-like' movement with robotic speed, dexterity, and repeatability. The robot is used for assembly, pick and place, machine tending, and other applications. These unique robots are designed with a central "torso" and two articulated arms. The Motoman SDA series has three robot models with graduating payload capacities: the SDA5D, SDA10D, and SDA20D. SDA robots have a total of 15 axes of movement, seven in each arm, one in the base. The two arms can be programmed to work independently or together. Motoman suggests that with a two arm robot, the need for costly positioners and tooling is minimized. One arm can act as the positioner, holding the part in place, while the other arm performs the rest of the application (i.e., welding, assembly) (see Figure 12 left). SDA robots also provide more payload and work envelope options than with single robot arms. When both arms work together, their payload capacity compounds. In the same manner, the two arms of SDA robots can work together to offer exceptional horizontal reach from side-to-side of the torso. Similarly, ABB (see Figure 14 right) has just unveiled their concept robot called FRIDA. They claim FRIDA's key technology features are:

- Harmless robotic coworker for industrial assembly
- Human-like arms and body with integrated controller
- Complements human labor with scalable automation
- Padded dual arms ensure safe productivity and flexibility
- Lightweight and easy to mount for fast deployment
- Agile motion

 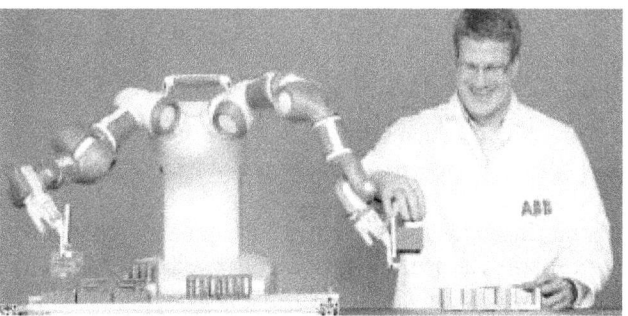

Figure 14 –Dual arm robots (left) [Motoman, 2010] and (right) [ABB, 2011]
[copyright photo-use permission granted by Motoman and ABB]

[Willow Garage, 2011] has developed the mobile PR2 robot with dual arms, torso, and Class 1 gripper as shown in Figure 15. PR2 is programmed using the ROS (robot operating system) open source robotic software framework and has a variety of features that provide potential for robotic assembly, including: back-driveable, current controlled, spring counterbalance arms; continuous two degrees of freedom wrists and enough torque to manipulate everyday objects from opening doors to handling frying pans; two grippers that can "grasp everything from towels to tea cups, and brooms to brews;" modular open interfaces to exchange different grippers, forearms, whole arms, or sensors; and low level and high level real time controllers with a software architecture for grasping and manipulation. Mobility includes a telescoping spine and an omni-directional base. Perception includes object recognition using lasers and cameras, open computer vision libraries, and 3D point cloud processing libraries.

Manufacturer claimed PR2 Manipulator Specifications:
    Arm Payload: 1.8 kg (4 lbs)
    Wrist Torque: 4 Nm (3 ft lbf)
    Grip Force: 80 N (18 lbf)

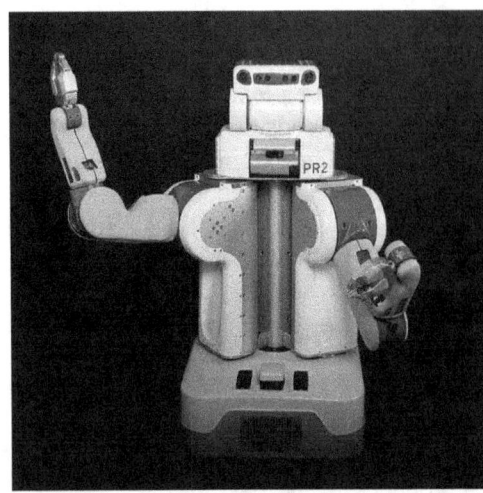

Figure 15 – PR2 Robot [Willow Garage, 2011, *shown is a publicly available image*]

b. Research

There is much research in advancing robotic assembly. This section divides the surveyed research into several areas including: touch, force, vision, threaded insertion, bin-picking, control, and U.S. Government research.

i. Touch

Surveying robot tactile or touch sensing uncovers most efforts from the 1980's and 1990's as shown in Section 2b(i) Tactile. Current examples of this research lie in tactile 'robot skin' systems.

Early research by [Inaba, 1996] presents the design and implementation of a tactile sensor suit that covers the entire body of a robot. The sensor suit, demonstrated on a full-body humanoid, was designed to be soft and flexible and to have a large number of sensing regions. They built the sensor suit using electrically conductive fabric and string. The current version of the sensor suit has 192 sensing regions. Each sensing region works as a binary switch in the current version. All of the signals from the sensor suit are gathered and superimposed on a visual image of the robot. A video multiplexer for the sensor signals is built on a field programmable gate array set.

[Kageyama, 1999] described two types of tactile sensor elements: a multi-valued touch sensor and a conductive gel sensor. The multi-valued touch sensor has multi-level pressure thresholds and is capable of covering wide areas of robot surfaces. The driving circuit is made as a module and scans the 128 switch elements in 1 ms. The switch elements have two pressure thresholds, with the first set to 25 gf/cm^2. The other sensor is made of conductive gel which has remarkable softness. This conductive gel has the advantage of its softness compared with other sheet type tactile sensors. The impedance of the gel changes by approximately 20 % from 0 gf/cm^2 to 400 gf/cm^2. The tactile sensor consists of 8×8 sensing points on the intersection of the electrodes, and the processing module scans all the points in 30 ms. Both sensors were again applied to a wheeled humanoid robot. [Kageyama, 1999]

More recently, [Hoshi, 2006] and [Barnard, 2010] developed a new tactile sensor skin ("skin by touch area receptor" or STAR). The skin consists of two components. One is a sensor element that detects a contact area in addition to a contact force. The element is inspired by the fact that humans can discriminate the sharpness of objects sensitively on any part of their bodies in spite of their several-centimeter two-point discrimination thresholds. The researchers developed the sensor element that has such characteristics in a very simple structure; two layers of compressible insulators (urethane foam) which are sandwiched between three pieces of stretchable conductive sheets (conductive fabric). The other component is a sensor/communication chip. The chips are arranged at the boundaries of the elements, and the chips measure the capacitances between the conductive layers and send signals through the same conductive layers. The chips enable connection of the elements to compose a soft robot skin including no long wires.

[Abhinav, 2009] describes a concept and presents experimental results for robotic end-effector operation when attached to an industrial robot for handling product items that are flexible or variable in physical properties such as size, shape, or firmness. This work also describes a scheme by which a manipulator can use dynamic tactile sensing to detect when it is about to lose hold of a grasped object and take preventive measures before gross sliding occurs. [Wikipedia (robotics), 2011] briefly discussed that scientists developed a prosthetic hand in 2009, called SmartHand, that functions like a real one - allowing patients to write, type on a keyboard, play piano, and perform other fine movements. The prosthesis has sensors that enable the patient to sense real feeling through the fingertips.

ii. Force

Assembly tasks involving large positional uncertainties are unsuitable for use of position-controlled robots. To automate such tasks, the assembly system must be responsive to contact forces. Issues in addressing force responsive automated assembly include contact stability, the degree of force responsiveness required for success, the speed of a successful implementation, and the means to program a force-responsive system to perform a given assembly task. Force control, or interaction control with the environment, can be performed with either indirect force control or direct force control. Indirect force control is motion control, without explicit closure of a force feedback loop. Direct force control offers the possibility of controlling the contact force to a desired value, thanks to the closure of a force feedback loop. [Siciliano, 1999]

[Newman, 1999] provides useful background references and information over several decades for force sensing used in assembly. The representative example of inserting a peg in a hole has been studied extensively. [Whitney, 1977] analyzes how to use a force feedback strategy to guide a simple cylindrical "peg-in-hole" assembly. [Whitney, 1982] discusses development of the quasi-static conditions to avoid wedging and jamming for the peg-in-hole. [Caine, 1989] provides analysis of non-chamfered peg-in-hole. [Raibert, 1981] developed controllers that used both position and force control for the manipulation of constrained objects. [Peshkin, 1990] then studied error correction and recovery during assembly. [McCarragher, 1994, 1995 and Austin, 1997] introduced synthesis methodologies involving a discrete event controller for a force-controlled system for assembly tasks. [Whitney, 1979] provided analysis of peg-in-hole assemblies at Draper Lab which led to the development of the remote-center compliance (RCC) wrist.

A remote center of compliance can be implemented as a passive mechanism in the wrist, instead of as an active control loop involving a force sensor, computer, and actuators. Using a passive RCC, a force-based strategy is encoded in the hardware in terms of the force vs. deflection characteristics. If the peg (and/or hole) is chamfered, the forces that arise due to small position errors cause the peg to self-align with the hole. Since the RCC is passive, there are no stability problems. This technique has demonstrated impressive assembly speeds. However, if the parts are not chamfered, or if the position uncertainties exceed the chamfer width, the RCC is ineffective. In general, if the force-guided strategy to be invoked cannot be realized in terms of a passive mechanism, then the RCC approach is inadequate. Caine, et. al., examined strategies where RCC for both non-chamfered, cylindrical and rectangular peg-in-hole insertion is not sufficient. They then developed strategies for constraining the allowable contact configurations between the parts in order to avoid configurations that will cause the assembly to fail.

Active algorithms for responding appropriately to sensed contact forces can accommodate more complex assembly cases and larger positioning uncertainties. In practice, these systems are too slow to be competitive with manual assembly. Attempts to speed up robots controlled with these algorithms have resulted in contact instability. [Zeng, 1997] reports on the existing robot force control algorithms and their composition based on the review of 75 papers on this subject. The objective was to provide a pragmatic exposition with specialty on their differences and different application conditions, and to give a guide to existing robot force control algorithms. The previous work can be categorized into discussion, design, and / or application of fundamental force control techniques, stability analysis of the various control algorithms, and advanced methods. Advanced methods combine the fundamental force control techniques with advanced control algorithms such as adaptive, robust and learning control strategies. While the majority of algorithms attempt reactive force control, [Gullapalli, 1994] presents a practical method for autonomous synthesis of appropriate admittance behavior for robust high-precision robotic assembly. They applied an on-line learning approach that relies on the appropriate admittance through repeated attempts at the assembly operation. They are able to circumvent the problems that alternative approaches have in trying to model the interactions between the robot and its environment. Their test results for the peg-in-hole insertion task show that the performance compares favorably with that of other proposed methods for high-precision, chamferless peg-in-hole insertion.

[Newman, 1999] also examined robotic assembly issues in the context of automotive transmission components and reported on an impedance-based, low-level algorithm and its interface to higher-level strategies that exhibited gentle, fast, and reliable assembly of example transmission components.

iii. Vision

Past research on industrial inspection and assembly suggests that problems in these areas are well suited to model-based analysis. The objects are man-made and manufactured from well-defined geometric descriptions. [Kruger, 1981] Visual feedback has traditionally been used in the assembly process to a very limited extent. With the advent of effective visual servoing techniques, visual feedback can become an integral part of the assembly process by complementing the use of force feedback to accomplish precision assemblies in imprecisely calibrated robotic assembly workcells. [Nelson, 1993]

Recent research from [Agrawal, 2009] presents a complete vision-guided robot system for model-based three-dimensional (3D) pose estimation and picking of singulated 3D objects. The system uses a novel vision sensor consisting of a video camera surrounded by eight flashes (light emitting diodes). By capturing images under different flashes and observing the shadows, depth edges or silhouettes in the scene are obtained. The silhouettes are segmented into different objects and each silhouette is matched across a database of object silhouettes in different poses to find the coarse 3D pose. The database is pre-computed using a computer-aided design (CAD) model of the object. The pose is refined using a fully projective formulation of Lowe's model-based pose estimation algorithm. The estimated pose is transferred to a robot coordinate system utilizing the hand–eye and camera calibration parameters, which allows the robot to pick the object. The authors suggest that the system outperforms conventional systems using two-dimensional sensors with intensity-based features as well as 3D sensors. They handle complex ambient illumination conditions, challenging specular backgrounds, diffuse as well as specular objects, and texture-less objects, on which traditional systems usually fail. The vision sensor is capable of computing depth edges in real-time and is low cost.

iv. Threaded Insertion

A step beyond peg-in-hole assembly is using self-tapping screws. Lara, et. al. introduced an automated robot-based system for the insertion of self-tapping screws into unthreaded holes. The system consists of three main components: a manipulator-guided screwdriver, a camera, and a system to control and monitor the overall process. They performed experiments in order to identify the requirements needed for a fully automated insertion system.

In robotics literature threaded insertion of screws in tapped holes is often referred to as a typical task; yet, unlike the smooth peg-in-hole problem, a robust control solution for inserting threaded fasteners has not been presented. [Nicolson, 1993] Relative screw versus nut orientation is a critical threading element. When errors are allowed only for the position along the axis of the bolt, the control problem for this restricted case is one dimensional. Therefore, there are two obstacles to the full automation of screw threading: 1) feeding and holding the parts, and

2) controlling the parts to ensure proper mating and to detect part failures in the presence of positional uncertainty. The main screw threading error is the inability to begin the thread when the bolt rotates too fast for a given axial spring constant and position for the spring equilibrium point. Nicolson, et. al. consider controlled compliance and accommodation matrix techniques claiming robust insertion of threaded fasteners. Errors in translational positioning are shown to be easily corrected. Errors in the angle of tilt between the threaded parts are shown to be much more difficult to correct and constrain the region of convergence for simple linear techniques.

[Diftler, 1999] discussed kinematic models that describe the relationship between threaded parts during back-spinning and determining angular alignment of bolts to nuts. They used a technique based on back-spinning a nut with respect to a bolt and measuring the force change that occurs when the bolt 'falls' into the nut.

v. Bin Picking

One robot manufacturer suggests that the future in autonomous robotic capability may be in random bin picking. Further, they suggest that there will be no need for unscrambling, collating, or datum positioning, and therefore, systems would become smaller, cheaper, and have far fewer elements. [ABB, 2010]

Research on machine vision and other sensor and software algorithm combinations have focused on bin picking due to its complexity. An internet search on 'bin picking' scholarly papers returned nearly 12,000 hits. [Hujazi,1990] suggests that prior research was based on intensity images and instead presents a segmentation algorithm using range images of industrial parts in a bin. A robot is then used for bin picking with a vacuum gripper making use of the algorithm to detect part edges and then perform region growing to build a final segmented image. [Ghita, 2003] says that generally, accurate 3-D information is required to develop versatile bin-picking systems capable of grasping and manipulation operations. After edge detection, as in the previous paper, Ghita suggests to recognize the object placed on the top of the object pile using a model-driven approach in which the segmented surfaces are compared with those stored in the model database. Finally, the attitude of the recognized object is evaluated using an eigen-image approach augmented with range data analysis. [Boehnke, 2007] used knowledge about the form of the objects to find them in range data, comparing the 2.5 dimension appearance of simulated object poses with the real range data. This approach takes features of range sensors into consideration to improve the accuracy and robustness of the object localization. [Boughorbel, 2003] suggested two types of sensors: range mapping scanners and video cameras for bin picking. The geometry of bin contents was reconstructed from range maps and modeled using superquadric representations, providing location and parts surface information that can be used to guide the robotic arm. Texture was also provided by the video streams and applied to the recovered models. The system is expected to improve the accuracy and efficiency of bin sorting and represents a step toward full automation.

vi. Control

Robot control for assembly is inherent within most sections of this paper, where appropriate part positioning and movement are required. This section surveys specific research being performed on algorithms for grasp, fuzzy control, and learning.

Grasp Algorithms

The complexity of the human hand, having 27 degrees-of-freedom, which on a conservative estimation can result in more than seven billion different hand poses, requires the derivation of simplified hand motions and hand postures. [Iberall, 1997] Basic types of hand prehension are shown in Figure 16 for different objects. [Shimoga, 1996] provides a survey of existing computational algorithms meant for achieving four important properties in autonomous, multi-fingered robotic hands, including: dexterity, equilibrium, stability, and dynamic behavior. Further, Shimoga suggests that multi-fingered robotic hands must be controlled so as to possess these properties and hence be able to autonomously perform complex tasks in a way similar to human hands.

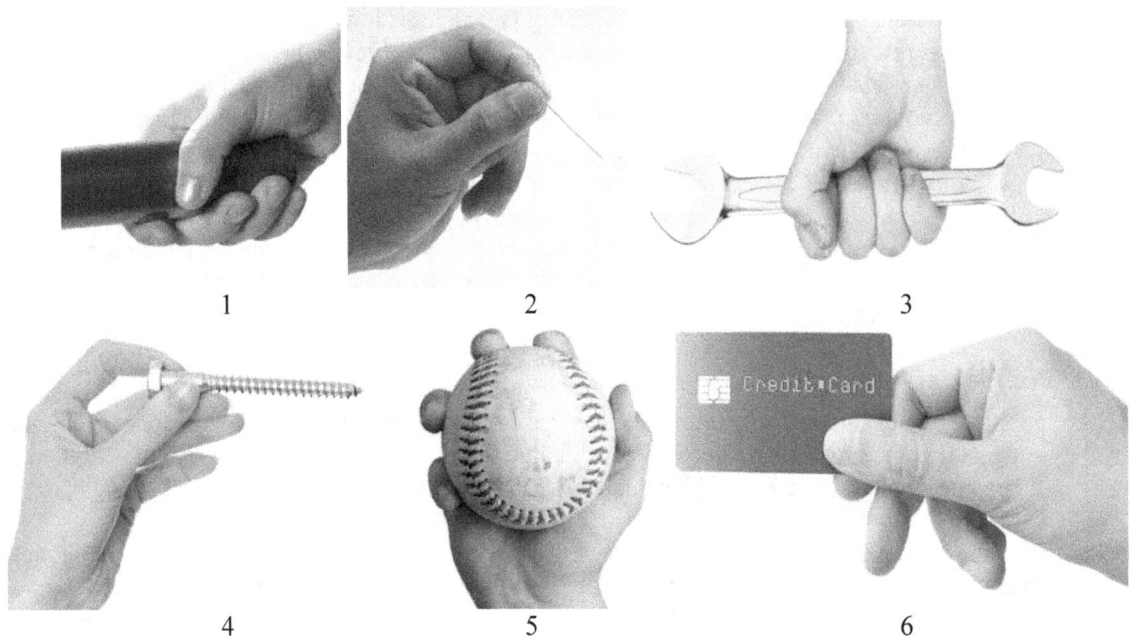

Figure 16 – Hand prehension of different objects:
1. Cylindrical hollow grip, 2. Tip grip, 3, Hook grip, 4. Three finger grip, 5. Hand palm grip, 6. Tong grip.
[The photos shown were approved for use by the following: ©Pete Saloutos/Shutterstock, ©discpicture/Shutterstock, ©Volodymyr Krasyuk/Shutterstock, ©April Cat/Shutterstock, ©GLYPHstock/Shutterstock, ©ilker canikligil/Shutterstock]

[Aiyama, 1998] designed compliance and planning motion of a 3-fingered gripper and manipulator control to handle a box in a compact array of six boxes as shown in Figure 17 (left). First one of the fingers tumbles one box to make two side faces free from the obstacles. Then the other two grasp both the sides and pick it up. To fulfill such a sequence of dexterous manipulations onto various objects, the design of compliance of the fingers is essential. Figure 17 (right) depicts a similar dexterous part manipulation with no grasping.

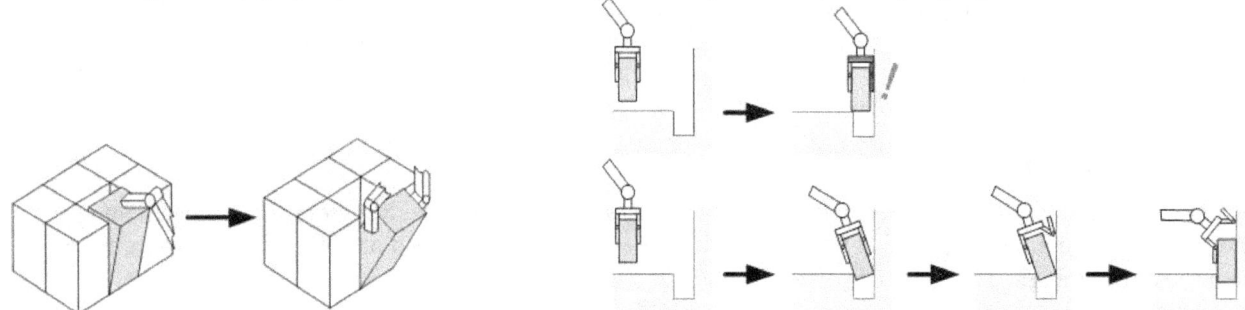

Figure 17 (left) dexterous manipulation and (right) part insertion by a non-grasping manipulation
[The graphics shown were redrawn by NIST based on images in Aiyama, 1998]

Fuzzy Control
[Soliman, 2009] explains their gripper designed to grasp unknown objects with different masses, shapes, and coefficients of frictions considering simplicity, durability, and economy. The grasping process during object lifting is considered mainly based on the slip reflex principle, as applying insufficient force leads to object slipping, and dropping may occur. A new fuzzy control algorithm based on empirical investigation of the human hand skills was proposed to adjust the applied force on the object without the risk of the object crushing or dropping. A simple rule base was used. Also, the controller was designed to maintain the object slip in reasonable

limits. The gripper design and developed force control algorithms resulted in the fast response of the task achievement. Input/output system variables are measured and analyzed. Experimental results obtained for different object masses and system disturbance show fast response in stopping the slippage and an enhancement in the grasping of different objects.

Control through Learning
[Gullapalli, 1994] describes their research in on-line learning for force control of peg-in-hole assembly, previously mentioned in Section 3b(ii) Advances in Robotic Assembly, Research, Force. The learning algorithm relies on the appropriate admittance through repeated attempts at the assembly operation.

[Kleinmanna, 2009] suggests that learning control systems are expected to have several advantages over conventional approaches when dealing with complex, high-dimensional processes. One example is the task of controlling grasp operations of a multi-fingered, multi-jointed robot gripper (the Darmstadt-Hand). The Advanced Gripper Control with Learning Algorithms (AGRICOLA) presented in this reference is able to maintain a stable grasp even if disturbances are applied. The algorithms also work for objects of different sizes for which the grasping has not been learned. Compared to the conventional stiffness approach, the performance of the learning system is equal, but the design is much easier since less knowledge about the gripper-hardware must be taken into account. The main part of the learning control loop is an associative memory that stores the grasping behavior as determined by the choice of an objective function.

[Bidaud, 1993] presents an advanced control system developed for an articulated gripper. This articulated gripper was previously designed to achieve stable grasp of objects with various shapes and to impart compliant fine motions to the grasped object. The researchers introduced autonomous reasoning capabilities in the control system of this device. Fine motion strategies, needed for mating or grasping, use inductive learning from experiments to achieve uncertainty and error recovery (on sensing, control, and modeling). The reference provides an overview of the articulated gripper's capabilities for a better understanding of the programming environment proposed. Declarative programming facilities in the controller were implemented for solving the problem of synthesis for fine motion planning through a time-sensitive expert system. Also, a heuristic procedure was used to obtain an implicit local model of contacts in complex assembly tasks.

vii. U.S. Government Research

[FRAPA, 1997] discusses a former NIST Advanced Technology Program (now Technology Innovation Program (TIP) [TIP, 2011]) joint project including industry, academia, and government for flexible robotic assembly of vehicle powertrain components. The project was approved because of the potentially large economic and technological gains, including:

- The total cost to U.S. industries of ergonomically related problems is in the range of $13 billion to $20 billion annually. Repetitive stress injuries (RSIs) are more likely to occur in assembly operations with heavy parts that require large mating forces.
- Powertrain and vehicular assembly have a high concentration of heavy parts that need to be manipulated repetitively without injury to the worker and without damage to the parts.
- Once autonomous powertrain component assembly is demonstrated in a flexible assembly work cell, the assembly technologies can be applied to numerous other component assemblies and applications, throughout various industries.

Figure 18 shows a graph of robotic application technology requirements and was the focus of the FRAPA flexible robotic assembly project. The result included a flexible system including a robot and vision system with requirements to be programmable and emulate human characteristics of force-controlled assembly using position control for large movements and changing over to force control before contacting parts. Also, the system included adaptive learning using artificial intelligence or genetic routines. A parallel axis robot resulted with natural admittance control and relatively high dexterity, including 15 µm precision, 200 Lb-f and 75 Lb nominal payload with a 1 m reach, and force sensing of 1 N (1/4 Lb).

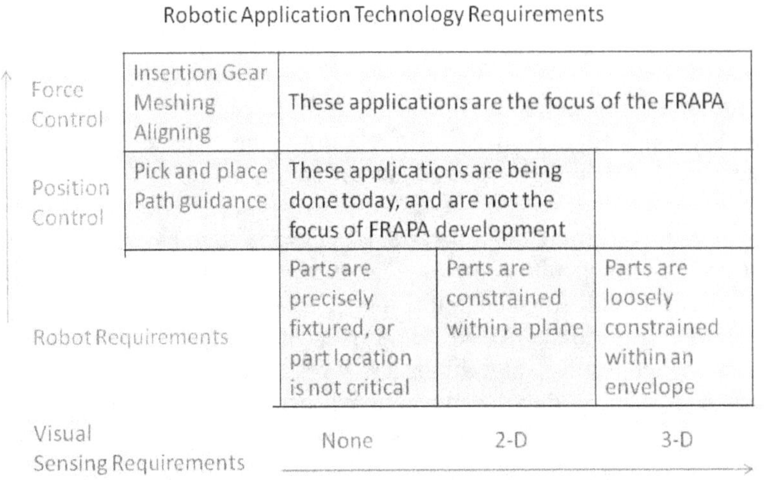

Robotic Application Technology Requirements

Force Control	Insertion Gear Meshing Aligning	These applications are the focus of the FRAPA		
Position Control	Pick and place Path guidance	These applications are being done today, and are not the focus of FRAPA development		
Robot Requirements		Parts are precisely fixtured, or part location is not critical	Parts are constrained within a plane	Parts are loosely constrained within an envelope
Visual Sensing Requirements		None	2-D	3-D

Figure 18 – Graph showing how the FRAPA project compared robot versus visual sensing requirements.

FRAPA included development of a 3D vision sensor with pose estimation and processing software that calculates the 6 DOF pose of parts in random orientations from true 3D vision data, and translates this data into robot coordinates, reducing part presentation and precision tooling requirements. The vision system trains itself on new parts, requiring no sophisticated programming or CAD inputs. The project team states that the FRAPA-developed vision system is the current state-of-the-art in direct 3-D image acquisition. All others are based on camera technology that do not return a true 3-D dataset and that rely on the inference of 3-D from 2-D data.

The Defense Advanced Research Projects Agency (DARPA) has been at the forefront of manipulation research. Three programs are currently underway: 1) Revolutionizing Prosthetics Program under the Biology Directive and Restorative Biomedical Technologies Thrust and managed by COL Geoffrey Ling, M.D., Ph.D., 2) Autonomous Robotic Manipulation (ARM) Program, and 3) Maximum Mobility and Manipulation (M3) Program under the Materials Directive and Multifunctional Materials and Material Systems Thrust managed by Gill A. Pratt, Ph.D. [DARPA, 2010]

The Revolutionizing Prosthetics Program will create, within this decade, a fully functional (motor and sensory) upper limb that responds to direct neural control. This revolution will occur by capitalizing on previous DARPA investments in neuroscience, robotics, sensors, power systems, and actuation. The program states that it will deliver a prosthetic for clinical trials that has function almost identical to a natural limb in terms of motor control and dexterity, sensory feedback (including proprioception), weight, and environmental resilience. The four-year device will be directly controlled by neural signals. The results of this program will allow upper limb amputees to have as normal a life as possible despite their severe injuries. Currently, prototypes from the two-year and four-year efforts are undergoing human testing.

The ARM program thrust is to enable autonomous manipulation systems to surpass the performance level of remote manipulation systems that are controlled directly by a human operator. ARM will create manipulators with a high degree of autonomy capable of serving multiple military purposes across a wide variety of application domains, including but not limited to counter-IED (improvised explosive device), countermine, search and rescue, weapons support, checkpoint and access control, explosive ordnance disposal, and combat casualty care (including battlefield extraction and treatment). The driver for this program is that current robotic manipulation systems save lives and reduce casualties, but still have many limitations. As examples, while manipulatation systems perform well in certain mission environments, they have yet to demonstrate proficiency and flexibility across multiple mission environments; they require burdensome human interaction and the full attention of the operator; and the time required to complete tasks generally exceeds military users' desires.

The DARPA Maximum Mobility and Manipulation (M3) program is striving to create and demonstrate significant scientific and engineering advances in robotics that will:

- Create a significantly improved scientific framework for the rapid design and fabrication of robot systems and greatly enhance robot mobility and manipulation in natural environments.
- Significantly improve robot capabilities through fundamentally new approaches to the engineering of better design tools, fabrication methods, and control algorithms. The M3 program covers scientific advancement across four tracks: design tools, fabrication methodologies, control methods, and technology demonstration prototypes.

c. Patents

Most patents shown below were found from an internet search on "robot gripper" and have U.S. Patent numbers of 7,xxx,xxx numbers or higher or with the new date/patent number designation from 2006 to present. As shown, the most recent inventions include special purpose gripping systems that build on previously demonstrated technology and are, in most cases, already in use. For example, multiple vacuum grippers as opposed to a single vacuum gripper, and grippers that are designed specific to part shapes and sizes, such as syringe and large jaw separation grippers.

U.S. Patent No. 7,690,706 B2, called: "Gripper Device," issued April 6, 2010:
Abstract: An apparatus for transporting objects may include a plurality of grippers having a first spacing at a first position and second spacing at a second position. A drive mechanism may be provided for selectively displacing the grippers from the first position to the second position to adjust the spacing between the grippers. See Figure 19.

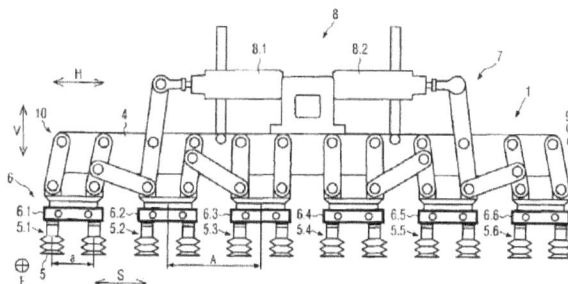

Figure 19 - "Gripper Device," from U.S. Patent No. 7,690,706 B2

U.S. Patent No. 7,134,833 B2, called: "Servo Adjustable Gripper Device," issued to Johannes J. M. de Koning, November 6, 2006:

Abstract: A servo adjustable gripper device for gripping and transporting at least two objects such as boxes includes a frame and at least two gripper assemblies connected with carriages slidably connected with the frame. The gripper device further includes a robotic arm which controls further movement and placement of the boxes on a pallet. The position of the gripper assemblies on the frame is adjustable to provide equal lifting force for a variety of differently sized boxes or for different numbers of boxes. See Figure 20.

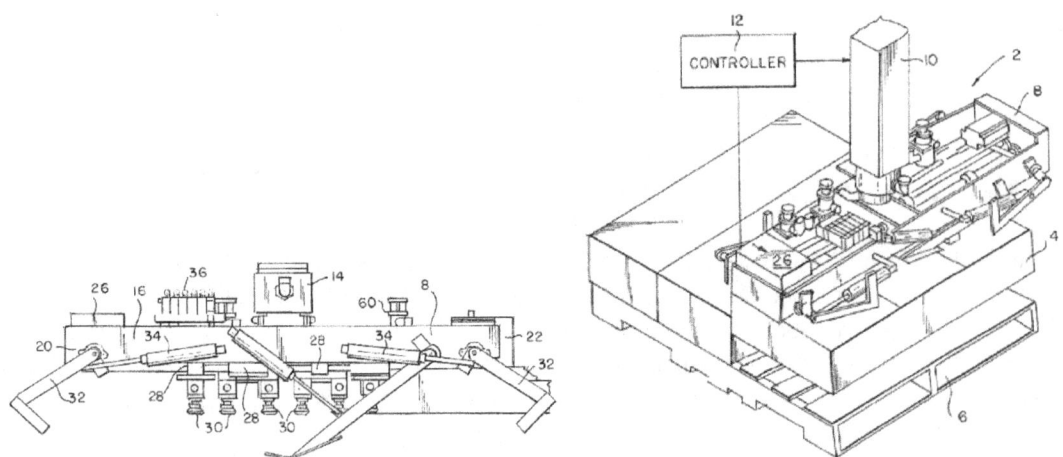

Figure 20 – (left) Side view and (right) isometric view of the "Servo Adjustable Gripper Device" shown in U.S. Patent No. 7,134,833 B2.

U.S. Patent No. 7,837,247 B2, called: "Gripper with Central Support," issued to Waldorf, Jenkins, and Kalb, November 23, 2010:

Abstract: A gripper assembly includes at least one gripper jaw and an actuator head linked with at least one gripper jaw. An actuator selectively operates to move the actuator head between a plurality of positions. A support is fixed relative to the actuator and includes a guide slot that guides the actuator head. One of the actuator heads or the guide slot includes a channel and the other of the actuator head or the guide slot includes a guide member extending at least partially into the channel. See Figure 21. A dual rod gripper version was also invented by Waldorf, et. al. under U.S. patent number 7854456.

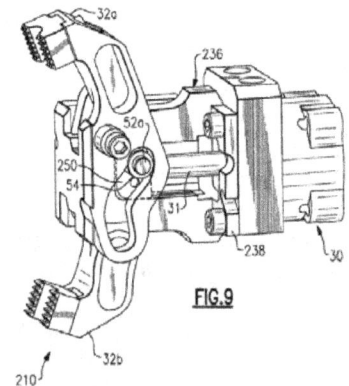

Figure 21 - "Gripper with Central Support" from U.S. Patent No. 7,837,247 B2.

U.S. Patent No. US 2009/0067973 A1, called: "Gripper Device," issued to Eliuk, Rob, Jones, and Deck, March 12, 2009:

Abstract: Gripper devices for handling syringes and automated pharmacy and mixture systems that utilize such gripper devices. The gripper devices may include various gripper finger profiles, substantially tapered or angled gripping surfaces, and/or gripper fingers interleaving to reduce radial distortion of the syringes to be grasped while opposing axial motion of the syringes. See Figure 22.

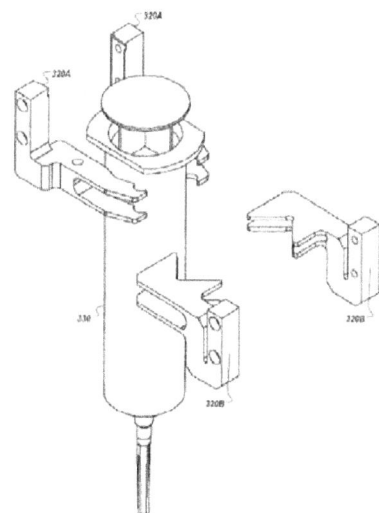

Figure 22 - "Gripper Device" from U.S. Patent No. US 2009/0067973 A1.

Other recent gripper patents found using the same search criteria as above include:

"Slide Gripper Assembly" having a slide assembly coupled to a gripper assembly. U.S. Patent No. 7,188,879 B2, by McIntosh, Steele, Givens, and Davenport on March 13, 2007

"Expandable Finger Gripper" for gripping the inside of containers or parts with cavities. U.S. patent no. 7,452,017 B2 by Maffeis on November 18, 2008.

"Quick Change Finger" that releasably connects a gripper finger to a robotic arm. U.S. patent no. 2010/0314895 A1 by Rizk and Delouis on December 16, 2010.

"Long Travel Gripper" for a rectilinear, large jaw separation gripper. U.S. Patent No. 7,490,881,B2 by Null and Williams on February 7, 2009.

"Automated Storage Library Gripper Apparatus and Method" for transporting and handling storage devices (cartridges). U.S. Patent No. 7,212,375 B2 by Dickey and Standt on May 1, 2007.

"Gripper System" having a pair of jaws and operates in one plane having a central closure axis. U.S. Patent No. US 2010/0164243 A1 by Albin on July 1, 2010.

"Stack Gripper" for gripping unbound printed products. U.S. Patent No. US 2007/0154292 A1 by Gammerler, Gunter, Meisel, Muller, Schubart on July 5, 2007.

4. Summary

This survey gave an overview of the current state of manipulation systems, as well as insight into future manipulation systems through the discussion of research being performed in the field. It included a methodical discussion of industrial manipulation capabilities, advancements, and research with a focus on autonomous assembly applications. The Appendices include a glossary of terms, a listing of industrial robot system standards, and a method of design for assembly.

In summary and with regard to end-of-arm tooling (EOAT) purchase, [EOAT, 1997] suggests the following tips that a user should follow:
- No one gripper style will secure every part.
- No matter how good the EOAT and no matter who built it, the user will always need to adjust it.
- Before buying any tooling, wise application-specific users compare the proposed EOAT with part drawings to ensure a good fit.
- The EOAT and the part weight together must not exceed the robot capacity. Also, choose tooling that's as light as possible to make the robot last longer.

This survey determined that the end-effector patents found are generally special purpose and that breakthroughs in highly capable assembly systems are minimal. As stated by [DARPA, 2011] with regards to defense robotics, "Robots hold great promise for amplifying human effectiveness in Defense operations. Compared to human beings and animals, however, the mobility and manipulation capability of present day robots is poor. In addition, design and manufacturing of current robotic systems are time consuming, and fabrication costs remain high. If these limitations were overcome, robots could assist in the execution of military operations far more effectively across a far greater range of missions." This statement can also be related to the capabilities of current robots needed for assembly operations. Dexterity, perception, and tactile capabilities of robot systems are advancing as exemplified by the multi-fingered grippers, the bin-picking capabilities, and the vision research. Also, safety and performance standards are beginning to consider collaboration of humans and robots within the same workspaces. However, several recommendations are listed in section 5 that suggest further advancements for robotic assembly systems.

5. Recommendations

A Smart Assembly Workshop [Smart, 2006] was held at NIST in 2006 to develop a vision and to define the state-of-the-art and industry needs in Smart Assembly. Smart Assembly refers to a next-generation capability in assembly systems and technologies which integrate "virtual" and "real time" methods in order to achieve dramatic improvements in productivity, lead-time, quality, and agility. The purpose of the workshop was to develop a broad industry/academic vision and to define the state-of-the-art and needs in "Smart Assembly."

Topics included:
- Defining and measuring aspects of smart assembly.
- Identifying key characteristics and attributes of smart assembly systems.
- Identifying critical scientific reserach challenges to enable smart assembly.
- Identifying critical implementation and infrastructure/standards challenges.
- Identifying models and opportunities for leveraging and collaboration to accelerate the development and implementation of smart assembly capability.

The workshop provided a substantial first step towards the formulation and launching of a Smart Assembly initiative. The characteristics and attributes for the future vision state were clearly defined, which fed the definition of priority recommendations, and suggested next steps were outlined towards a unified program. One specific recommendation was that NIST should consider the creation of a National Smart Assembly Testbed in cooperation with industry sponsors to validate the interoperability and performance of smart assembly modules and systems.

The following are recommendations from literature and from interviews with manufacturers. These recommendations may be useful for product development, research planning, and standards development.
- Develop performance metrics for autonomous assembly - current specifications from vendor products are limited to accuracy and resolution. Performance measurements should include both robot system performance measurements (i.e., overall system performance) and system component performance measurements (e.g., force sensor, gripper, robot).

- o An example performance measure testbed is shown in figure 23. The figure depicts an independent test setup that incorporates a 6-axis load cell to measure applied forces of an assembly operation, as well as definable measurements of success. The four step assembly shows a set of spur gears to be assembled using various standard force control capabilities and the resultant forces are monitored throughout the process using an independent load cell or force/torque sensor.
- o Other examples are: standard peg-in-hole tests (smooth peg or screws), slides, etc.
- o Vision type assembly tasks
- o Associated with these tests are, for example, force and other sensors with stock vendor algorithms having potentially many tuning variables

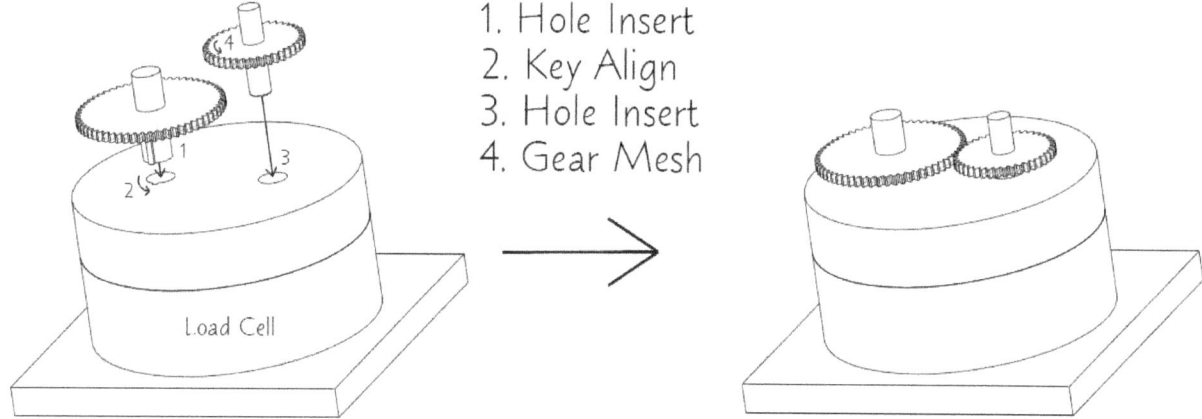

Figure 23 - NIST Performance Measures for Assembly concept drawing

[Schofield, 2010] stated that:
- Increased use of robots as 'intelligent' fixturing will make systems more flexible and give faster product change over times
- External metrology will open up many applications where robots have previously not been sufficiently accurate
- More human/robot interaction in a safe environment will maximize productivity
- There is a need for reduced sizes of end-effector devices – compliance, force/torque sensors, quick changes, sensors (proximity, tactile)
- There is a need to match autonomous robot systems capabilities to humans, i.e., as compared to precision tactility and dexterity by a surgeon, watchmaker, or jewelry maker.
    - o tactility – there are limited/no sight applications where the robot needs to 'feel' to perform high precision assembly (e.g., start an M3 screw or meshing small gears, parts)
    - o dexterity – manipulating tools (e.g., tweezer sized grippers) and parts (e.g., pins, rivots) for precision assembly

Recommendations from interviews with end-effector and robot manufacturers at the ProMat and Automate 2011 events in Chicago, IL, March 2011 who preferred to remain anonymous are as follows:

End-Effectors:
- Clear, standard interfaces for grippers
- Integration of current industrial grippers with sensors to sense that a part being gripped has been acquired.
- Verification of perception systems with low-cost gripper tactile sensors
- Adaptive end-effectors so that robot control includes gripper control - for example, unlike typical human arm/hand control, current robot and gripper systems allow the robot to move to a part, then the gripper grasps

the part, and the robot moves the part. Alternatively, this motion would be done simultaneously with gripper feedback to the robot controller to adjust the arm and grip during the entire part acquisition process. Low cost tactile sensing for off-the-shelf pneumatic grippers is needed.

Robots:
- "Feel your way through the assembly process" including robots supported by low mass/quick response, high update rates, direct impedance control per joint.
- Hand guiding robots through the teaching process.
- Path planning of two-arm robots so they don't interfere or collide with each other.
- Robots on vehicles - The issue is that the automated guided vehicle (AGV) is addressed by standards maintained by the industrial truck interest group. This situation was great for the initial applications of the load-carrying AGV (based on identifiable tracks in an industrial setting), but has become problematic as the AGV is adapted to become more of a co-worker to the human. Since the redefinition of the term 'robot' within the International Organization for Standardization (ISO), the AGV now fits within its purview. The inclusion of roving robot arms and similar applications in a holistic structure is not currently being considered by industrial standards organizations.
- Dynamic work volumes, for example robot arms on vehicles, that are intelligently controlled to avoid obstacles as the vehicle moves the robot and its payload. Included here is intelligent environment sensing and control.
- Tactile response is critical in assembly operations, and has been most difficult to develop with robots. A robot does an assigned task very well and repetitively, but has problems with variable situations.
- Collaborative robots, since they will play a huge part in future assembly applications. More work is needed in this area. The ISO/TS 15066 technical specification for how to implement robot collaboration with humans and other equipment (e.g., AGV's) will address some of this need. The ISO 10218-1 standard provides the manufacturer with information for construction of collaborative robots. The ISO 10218-2 standard provides the integrator and user guidance on how to use collaborative robots.
- Measurement methods are needed for the component inaccuracies stemming from the robot, end-effector, sensors, fixturing, dynamics, joint flex, or other inputs to the assembly process, so that robots can be designed to meet high accuracy, assembly applications.

An example scenario suggested by robot manufacturers was to assemble a kit of parts, such as a disposable camera, having small parts that mesh and require robot control, including force sensing, perception, and high dexterity. During an industry workshop on Dynamic Perception [Eastman, 2009], attendees discussed the scenario of assembling a kit for the purpose of measuring robot assembly system performance:

"Kit loading and unloading has the advantage of offering well-defined, transportable test artifacts. A manufacturing assembly challenge might consist of a suitcase-style case that folds open with hinges that allow the two sides to be disconnected. In one half would be a set of parts of specified sizes and shapes in holders designed for each. The other half would be a bin. There could also be a mat with outlined spots for each part. The challenge kit would support a number of tasks. All the parts could be dumped in the bin side and then picked and placed onto the kit side in the designated locations. Alternatively, the kit side could be unloaded part by part into the bin. For evaluating robot precision in grasping, the kit side could be unloaded onto the mat to match the outlines (there are no supporting sides to guide a part.) The parts could be varied to provide both a recognition task and an orientation and gripping task, and the parts could vary in difficulty of recognition and gripping. For mobile manipulators, the bin and kit could be separated. The parts could be designed with different geometries, such as prismatic, cylindrical, or ovoid, so that some are easier for pose calculations and some are harder. For single-arm robot systems, the assembly should either have, or come with, a stable base to build on. Beyond simple parts handling, the kit could be assembled with or without human help. The parts could be designed for various assembly operations, such as peg-in-hole with gravity to hold them in place, screw threads, etc., and with different levels of difficulty. Increased dimensional tolerance could intentionally be added to the base parts to make assembly easier, but also to prevent the robot from using dead reckoning for assembly. Over time, the kits could be made more complex."

6. References

[ABB, 2005] New ABB wireless sensor takes machine control to the next level, http://www.abb.com/cawp/seitp202/7cee4f52c2174d6180257006002c5227.aspx, May, 19.

[ABB, 2006] Application Manual, Force Control for Assembly, Controller Software IRC5, RobotWare 5.0, Document ID: 3HAC 025057-001, Revision: B, SE-721 68. Västerås, Sweden.

[ABB, 2010] Ian Schofield, A perspective on the characteristics of Advanced Industrial Robotic Systems presentation, October 22.

[ABB, 2011] FRIDA dual arm robot, http://www.abb.com/cawp/abbzh254/fcfbdad9a72cfe08c1257862006bcfbf.aspx, April 6.

[Abhinav, 2009] Abhinav V and Vivekanandan S, Real-Time Intelligent Gripping System for Dexterous Manipulation of Industrial Robots, Proceedings of the World Congress on Engineering 2009 Vol II WCE 2009, London, U.K, July 1 – 3.

[Agrawal, 2009] Amit Agrawal, Vision-guided Robot System for Picking Objects by Casting Shadows, Yu Sun, John Barnwell, Ramesh Raskar, International Journal of Robotics Research, Vol. 1, No. 29, November 13.

[Aiyama, 1998] Y. Aiyama, T. Arai and J. Ota, Dexterous Assembly Manipulation of a Compact Array of Objects, CIRP Annals - Manufacturing Technology, Volume 47, Issue 1, Pages 13-16.

[AMI, 2011] Design for Assembly analysis, http://www.ami.ac.uk/courses/ami4945_dpb/restricted/u08/supplementary/sup_04.html.

[ANSI, 2011] American National Standards Institute, www.ansi.org.

[Assembly, 2005] Design for Robotic Assembly article, www.assemblymag.com, September 20.
[ASTM, 2011] ASTM International Standards, www.astm.org.

[ATI, 2007] ABB introduces new robot control system for automating assembly tasks; RobotWare Assembly FC utilizes ATI's Force/Torque Sensors, http://www.ati-ia.com/company/NewsArticle.aspx?id=371669795, February 22.

[Austin, 1997] D. Austin and B. J. McCarragher, Force Control Command Synthesis for Assembly Using a Discrete Event Framework. Proceedings of 1997 IEEE International Conference on Robotics and Automation, Albuquerque, NM, pp. 933 - 938, April, 1997.

[Barnard, 2010] Patrick Barnard, Peratech Commissioned to Develop 'Tactile Robotic Skin' that Allows Robots to 'Feel' Touches. February 24.

[Bidaud, 1993] P. Bidaud and D. Fontaine, A learning control system for an articulated gripper, Experimental Robotics II, Lecture Notes in Control and Information Sciences, Volume 190/1993, 99-111, DOI: 10.1007/BFb0036134.

[Boehnke, 2007] Boehnke, K., Object localization in range data for robotic bin picking, IEEE International Conference on Automation Science and Engineering, CASE 2007, September 22-25.

[Bostelman, 1989] Roger Bostelman, Electronics Design of the Infrared/Ultrasonic Sensing for a Robot Gripper, NISTIR 89-4223, National Institute of Standards and Technology, November.

[Boughorbel, 2003] Faysal Boughorbel, Yan Zhang, Sangkyu Kang, Umayal Chidambaram, Besma Abidi, Andreas Koschan, Mongi Abidi, "Laser ranging and video imaging for bin picking", Assembly Automation, Vol. 23 Iss: 1, pp.53 – 59.

[Brumson, 2011] Bennett Brumson, After 50 Years Robots Have New Horizons Robotic Industries Association, www.robotics.org, January 11.

[Bruno, 2011] Leonard C. Bruno, Assembly Line - Role Of Workers, http://science.jrank.org/pages/559/Assembly-Line-Role-workers.html.

[Caine, 1989] Caine, Michael E., Tomás Lozano-Pérez, and Warren P. Seering. *Assembly Strategies for Chamferless Parts*. Proceedings of the IEEE International Conference on Robotics and Automation, Pp. 472-477.

[Cho, 1987] Cho, H.S., Warnecke, H.J., Gweon D.G., Robotic assembly: a synthesizing overview, Robotica 5: 153-165, Cambridge University Press.

[Cornell, 2010] Ju, Anne, Balloon filled with ground coffee makes ideal robotic gripper, http://www.news.cornell.edu/stories/Oct10/UniversalGripper.html, Chronicle Online, October 25.

[Crowder, 1998] Crowder, RM, Automation and Robotics: Tactile Sensing, http://www.soton.ac.uk/~rmc1/robotics/artactile.htm.

[Cutkosky, 1993] Mark R. Cutkosky, James M. Hyde, 6th ISRR, Manipulation Control with Dynamic Tactile Sensing, Hidden Valley, Pennsylvania, Oct. 2-5.

[DARPA, 2011] Defense Advanced Research Project Agency Programs, www.darpa.mil/dso/thrusts

[Delden, 2006] Sebastian van Delden, Constructing a Simple Visually-Guided Robotic Part-Grasping System with Off-the-Shelf Components, Proceedings of the 18th IEEE International Conference on Tools with Artificial Intelligence (ICTAI'06)

[DeSouza, 2004] Guilherme N. DeSouza and Avinash C. Kak, " A Subsumptive, Hierarchical, and Distributed Vision-Based Architecture for Smart Robotics ," IEEE Transactions on Systems, Man, and Cybernetics -- Part B: Cybernetics, Vol. 34, pp. 1988-2002, October 2004

[Diftler, 1999] Diftler, M.A., Walker, I.D., Experiments in aligning threaded parts using a robot hand, IEEE Transactions on Robotics and Automation, Vol. 15, Issue 5, Pp, 858-868, October.

[Draper, 2008] Tom Draper, How to Use Color Sensors to Improve Product Quality, Design World Magazine, http://www.balluff.com/Balluff/us/NewsChannel/Articles/en/2008-06_Color+Sensor.htm, June.

[Dictionary, 2010] Online Dictionary, www.thefreedictionary.com.

[DLR, 2011] Multi-sensor five-fingered hand with fifteen degrees of freedom http://www.dlr.de/rm/en/desktopdefault.aspx/tabid-4789/7945_read-12721/.

[Eastman, 2009] Roger Eastman, Elena Messina, Tsai Hong, et. al., Dynamic Perception Workshop Report, Requirements and Standards for Advanced Manufacturing, Rosemont, Illinois, June 11.

[EOAT, 1997] Fisher, Trent, Get a Grip on the Basics of Robot End-of-Arm Tooling, http://www.eoat.com/articles/SAS_US/article.htm.

[Fanuc, 2007] Fanuc Robot Series, Force Sensor Operator's Manual, # B-81154EN/03, Revised March.

[Fanuc, 2011] Fanuc Robotics, http://www.fanucrobotics.com/products/vision-software/AtoZ.aspx.

[FRAPA, 1997] Motor Vehicle Manufacturing Technology (October 1997), Flexible Robotic Assembly for Powertrain Applications (FRAPA), http://jazz.nist.gov/atpcf/prjbriefs/prjbrief.cfm?ProjectNumber=97-02-0018.

[Genaldy, 1990] A. M. Genaldy, J. S. Duggal, A. Mital, A comparison of robot and human performances for simple assembly tasks, International Journal of Industrial Ergonomics, Volume 5, Issue 1, Pages 73-81, January.

[Ghita, 2003] Ovidiu Ghita, Paul F. Whelan, A bin picking system based on depth from defocus, Machine Vision and Applications, 13: 234–244.

[Gullapalli, 1994] Gullapalli, Vijaykumar, Andrew G. Barto, and Roderic A. Grupen. Learning Admittance Mappings for Force-Guided Assembly. Proceedings of the IEEE International Conference on Robotics and Automation, Pp. 2633-2638.

[Hardin, 2005] Winn Hardin, 3D vision: Simpler, Smarter, Spreading, Automated Imaging Association, June 8. http://www.machinevisiononline.org/vision-resources-details.cfm/vision-resources/3D-vision-Simpler-Smarter-Spreading/

[Hoshi, 2006] Hoshi, T., Shinoda, H., Robot skin based on touch-area-sensitive tactile element, Proceedings of the IEEE International Conference on Robotics and Automation, (ICRA), May 15-19.

[Howe, 1994] Robert, D. Howe, Tactile Sensing and Control of Robotic Manipulation, Journal of Advanced Robotics, Volume 8, Issue 3, Pages 245-261.

[Hujazi, 1990] Al-Hujazi, E | Sood, A, Range image segmentation with applications to robot bin-picking using vacuum gripper, IEEE Transactions on Systems, Man, and Cybernetics, Vol. 20, no. 6, pp. 1313-1325.

[Iberall, 1997] Thea Iberall, Human Prehension and Dexterous Robot Hands, International Journal of Robotics Research.

[Inaba, 1996] Inaba, M.; Hoshino, Y.; Nagasaka, K.; Ninomiya, T.; Kagami, S.; Inoue, H., '96 Proceedings of the Intelligent Robots and Systems (IROS), November 4-8.

[ISO, 2011] International Organization for Standardization, www.iso.org.

[Iversen, 2006] Wes Iversen, Vision-guided Robotics: In Search of the Holy Grail, Automation World, February.

[Kageyama, 1999] Kageyama, R., Kagami, S., Inaba, M., Inoue, H., Development of soft and distributed tactile sensors and the application to a humanoid robot, Proceedings of the IEEE International Conference on Systems, Man, and Cybernetics, October 12-15.

[Kleinmanna, 2009] K. Kleinmanna, M. Hormela and W. Paetscha, Intelligent real-time control of a multifingered robot gripper by learning incremental actions, Darmstadt University of Technology, Department of Control Systems Theory and Robotics, Darmstadt, Germany, 8 October

[Kruger, 1981] Kruger, Richard P., Thompson, William B., A Technical and Economic Assessment of Computer Vision for Industrial Inspection and Robotic Assembly, Proceedings of the IEEE, Vol. 69, No. 12, December.

[Kuka, 2010] Handling of automotive safety parts, http://www.kuka-robotics.com/germany/en/solutions/solutions_search/L_R239_Handling_of_automotive_safety_parts.htm

[Lara, 1998] Bruno Lara, Kaspar Althoefer, Lakmal D. Seneviratne, Automated Robot-based Screw Insertion System, IECON '98. Proceedings of the 24th Annual Conference of the IEEE Industrial Electronics Society.

[Masi, 2006] Masi, C.G., *Machine Vision: Not Just For Metrology Anymore*, Control Engineering, www.controleng.com, October 1.

[Maitland, 2008] Murray E. Maitland, Molly Epstein, Analysis of Finger Position During Two and Three-Fingered Grasp: Possible Implications for Terminal Device Design, MEC.

[McCarragher, 1994] B. J. McCarragher, Error Detection and Recovery of Robotic Assembly Tasks. IFAC Symp. Robot Control, pp. 891-896, September.

[McCarragher, 1995] B. J. McCarragher and H. Asada, The Discrete Event Modeling and Trajectory Planning of Robotic Assembly Tasks. ASME Journal of Dynamic Systems,
Measurement and Control, Vol. 117, No. 3, pp. 394 - 400, Sept. 1995.

[Mindtrans, 2010] "The best anthropomorphic robot hands/arms," http://mindtrans.narod.ru/hands/hands.htm. July.

[Moon, 2009] Seungbin Moon, Gurvinder S. Virk, Survey on ISO Standards for Industrial and Service Robots, ICROS-SICE International Joint Conference 2009, Fukuoka International Congress Center, Japan, August 18-21.

[Monkman, 2007] Gareth J. Monkman, Stefan Hesse, Ralf Steinmann, Henrik Schunk, Robot Grippers, Wiley-VCH.

[Motoman, 2010] The Motoman SDA Two Arm Robot Series http://www.robots.com/blog.php?tag=458, September 17.

[MMPMS, 2011] Mobility and Manipulation Performance Measurements and Standards Project, National Institute of Standards and Technology, http://www.nist.gov/el/isd/ms/mmp.cfm.

[Nelson, 1993] Brad Nelson, , N.P., Papanikolopoulos, P.K. Khosla, Visual Servoing for Robotic Assembly, *Visual Servoing-Real-Time Control of Robot Manipulators Based on Visual Sensory Feedback*, ed. K. Hashimoto, World Scientific Publishing Co. Pte. Ltd., River Edge, NJ, pp. 139-164.

[Newman, 1999] Wyatt S. Newman, Michael S. Branicky, H. Andy Podgurski, Siddharth Chhatpar, Ling Huang, Jayendran Swaminathan, Hao Zhang, Force-Responsive Robotic Assembly of Transmission Components, Proceedings of the 1999 IEEE International Conference on Robotics & Automation, Detroit, Michigan, May.

[Nof, 1999] Shimon Y. Nof, Handbook of Industrial Robotics, Editor, 2nd Edition, 1999

[Nicolson, 1993] Edward J. Nicolson and Ronald S. Fearing, Compliant Control of Threaded Fastener Insertion, 2002-IEEEexplore.IEEE.org.

[Ogando, 2007] Ogando, Joseph, Force Control and Machine Vision Guide Robots, Design News, June 24.

[Packworld, 2009] Contract packager crafts portable, flexible solution, Robots and Automation, http://www.packworld.com/article-27899, Packaging World Magazine, p. 42, August.

[Paulos, 1993] Eric Paulos John Canny, "Informed peg-in-hole insertion using optical sensors" SPIE Conference on Sensor Fusion VI. Boston Massachusetts.

[Perry, 2002] Dwayne Perry, Force/torque sensors help industrial robots make the right moves, Machine Design, Edited by Lawrence Kren, January 24.

[Peshkin, 1990] M. A. Peshkin, Programmed Compliance for Error Corrective Assembly. IEEE Transactions on Robotics and Automation, Vol. 6, No. 4, pp. 473 - 482, August.

[Picard, 2002] Marie-Pierre Picard, FlexPlace: Watchmaker precision for robotic placement of automobile body parts, Industrial Robot Journal, Vol., 29, No. 4, pp. 329-333.

[Prahlad, 2011] Harsha Prahlad, presentation to NIST Intelligent Systems Division, http://www.hizook.com/blog/2010/10/28/electroadhesive-robot-grippers-sri-international, March.
[Raibert, 1981] M. Raibert, H. Mason, and J. J. Craig, Hybrid Position/Force Control of Manipulators. ASME Journal of Dynamic Systems, Measurement, and Control, Vol. 102.

[Pressure Profile, 2010] Tactile Pressure Measurement Systems, http://www.pressureprofile.com/products.php.

[RoMeLa, 2009] *Laura June,* Robotic hand controlled by compressed air grasps the concept of delicacy, Engadget, http://www.engadget.com/2009/05/07/robotic-hand-controlled-by-compressed-air-grasps-the-concept-of/, May 7.

[Schofield, 2010] Ian Schofield, Advanced Industrial Robotic Systems, ABB Group Presentation PDF, October 22.

[Shafi, 2007] Adil Shafi, How to Implement Bin Picking in your Manufacturing Operation, Robotics Online, April 4.

[Shimoga, 1996] K. B. Shimoga, Robot Grasp Synthesis Algorithms: A Survey, International Journal of Robotics Research, Vol. 15, No. 3, PP. 230-266, June.

[Siciliano, 1999] Bruno Siciliano, Luigi Villani, Robot force control, Kluwer Academic Publishers.

[Smart, 2006] Smart Assembly: Industry Needs and Technical Challenges, NIST Workshop on Smart Assembly, http://smartassembly.wikispaces.com/, October 3-4.

[Soliman,2009] A. M. Soliman, A. M. Zaki, A.M. El-Shafei, O. A. Mahgoub, A Robotic Gripper Based on Advanced System Set-up and Fuzzy Control Algorithm, Proceedings of the IEEE International Conference on Automation and Logistics Shenyang, China, August.

[TIP, 2011] NIST Technology Innovation Program website, http://www.nist.gov/tip/.

[Volpe, 1994] Richard Volpe, Robert Ivlev, A Survey and Experimental Evaluation of Proximity Sensors for Space Robotics, proceedings of The IEEE International Conference on Robotics and Automation, May.

[Sivam, 2004] Sivam E-Publishing, Welding Technology Machines, http://www.welding-technology-machines.info.

[Whitney, 1977] D. E. Whitney, Force Feedback Control of Manipulator Fine Motions. ASME Journal of Dynamic Systems,Measurement, and Control, pp. 91 -97, June.

[Whitney, 1979] D. E. Whitney and J. L. Nevins, What is the Remote Center Compliance (RCC) and What Can It Do?, Proceedings of the 9th International Symposium on Industrial Robots, Washington, DC, March.

[Whitney, 1982] D. E. Whitney, Quasi-static Assembly of Compliantly Supported Rigid Parts. ASME Journal of Dynamic Systems, Measurement and Control.

[Wikipedia, 2011] www.wikipedia.com.

[Willow Garage, 2011] http://www.willowgarage.com/pages/pr2/overview.

[Yaskawa, 2011], PackWorld, http://motoman.com/products/integrated/packworld.php

[Youngrock, 2008] Youngrock Yoon, Akio Kosaka, Avinash C. Kak, A New Kalman-Filter-Based Framework for Fast and Accurate Visual Tracking of Rigid Objects, IEEE Transactions on Robotics, Vol. 24, No. 5, October.

[Zeng, 1997] Zeng, Ganwen, and Ahmad Hemani. An *Overview of Robot Force Control*. Robotica, Vol. 15. Pp. 473-482.

[Zenzen, 2001] Zenzen, Joan M., Automating the Future - A History of the Automated Manufacturing Research Facility 1980-1995, Diane Publishing Company, Paperback Book, 98 pages, March 01.

[Yoon, 2008] Youngrock Yoon, Akio Kosaka, Avinash C Kak, "A New Kalman-Filter-Based Framework for Fast and Accurate Visual Tracking of Rigid Objects," IEEE Transactions on Robotics, vol. 24, No. 5, October 2008

7. Appendix

a. Terminology

This terminology was extracted from [Monkman, 2007] and [Nof, 1999]. The source for each term below is designated as either [1] or [2] respectively. The terminology extracted was chosen based on how well the definition fit the context of this document. In some cases, there are duplicate definitions for the same term.

Accuracy[2]: The ability of a robot to position its end-effector at a programmed location in space. Accuracy is characterized by the difference between the position to which the robot tool-point automatically goes and the originally taught position, particularly at nominal load and normal operating temperature.

Actuator[2]: A motor or transducer that converts electrical, hydraulic, or pneumatic energy into power for motion or reaction.

Adaptive Control[2]: A control method used to achieve near-optimum performance by continuously and automatically adjusting control parameters in response to measured process variables. Its operation is in the conventional manner of a machine tool or robot with two additional components:

1. At least one sensor which is able to measure working conditions; and
2. A computer algorithm which processed the sensor information and sends suitable signals to correct the operation of the conventional system

Artificial Intelligence (AI)[2]: The ability of a machine system to perceive anticipated or unanticipated new conditions, decide what actions must be performed under the conditions, and plan the actions accordingly.

Assembly (Robotic)[2]: Robot manipulation of components resulting in a finished assembled product.

Assembly Constraints[2]: Logical conditions that determine the set of all feasible assembly sequences for a given product. Assembly constraints can be of two types: geometric precedence constraints (those arising from the part geometry) and process constraints (those arising from assembly process issues).

Astrictive gripper[1]: A binding force produced by a field is astrictive. This field may take the form of air movement (vacuum suction), magnetism or electrostatic charge displacement.

Automation[2]: Automatically controlled operation of an apparatus, process, or system by mechanical or electronic devices that replace human observation, effort, and decision.

Basic jaw (universal jaw)[1]: The part of an impactive gripper subjected to movement. An integral part of the gripper mechanics, the basic jaw is not usually replaceable. However, the basic jaws may be fitted with additional fingers in accordance with specific requirements.

Basic unit[1]: Basic module containing all gripper components which is equipped for connecting (flange, hole pattern) the gripper to the manipulator. The connecting capability implies a mechanical, power, and information interface.

Chemoadhesion[1]: Contigutive prehension force by means of chemical effects (usually in the form of an adhesive).

Closed –Loop Control[2]: The use of a feedback loop to measure and compare actual system performance with desired performance. This strategy allows the robot control to make any necessary adjustments.

Compliance[2]: A feature of a robot which allows for mechanical float in the tooling in relation to the robot tool mounting plate. This feature enables the correction of misalignment errors encountered when parts are mated during assembly operations or loaded into tight-fitting fixture or periphery equipment.

Compliant Assembly[2]: The deliberate placement of a known, engineered, and relatively large compliance into tooling in order to avoid wedging and jamming during rigid part assembly.

Contigutive Gripper[1]: Contigutive means touching. Grippers whose surface must make direct contact with the objects surface in order to produce prehension are termed contigutive. Examples include chemical and thermal adhesion.

Compliant Support[2]: In rigid part assembly, compliant support provides both lateral and angular compliance for at least one of the mating parts.

Contact Sensor[2]: A grouping of sensors consisting of tactile, touch, and force/torque sensors. A contact sensor is used to detect contact of the robot hand with external objects.

Control System (Gripper)[1]: In most of the cases a relatively simple control component for analyzing of pre-processing sensor information for regulation and/or automatic adjustment of prehension forces.

Conveyor Tracking Robot[2]: A robot synchronized with the movement of a conveyor. Frequent updating of the input signal of the desired position on the conveyor is required.

Degrees of Freedom[2]: The number of independent ways the end-effector can move. It is defined by the number of rotational or translational axes through which motion can be obtained. Every variable representing a degree of freedom must be specified if the physical state of the manipulator is to be completely defined.

Dextrous hand[1]: Anthropoidal artificial hand (rarely for industrial use), which is equipped with three or more jointed fingers and may be capable of sophisticated, programmed or remote controlled operations.

Disassembly[2]: the inverse of the assembly process, in which products are decomposed into parts and subassemblies. In product remanufacturing the disassembly path and the termination goal are not necessarily fixed, but rather are adapted according to the actual product condition.

Double grippers[1]: Two grippers mounted on the same substrate, intended for the temporal and functional prehension of two objects independently.

Drive system[1]: A component assembly which transforms the applied (electrical, pneumatic, hydraulic) energy into rotary or translational motion in a given kinematic system.

Dual Grippers[1]: Tow grippers mounted on the same substrate, intended for the simultaneous prehension of two objects.

Electroadhesion[1]: prehension force by means of an electrostatic field.

End-Effector[1] [2]: Also known as end-of-arm tooling or, more simply, hand.

Generic term for all functional units involved in direct interaction of the robot system with the environment or with a given object. These include grippers, robot tools, inspection equipment and other parts at the end of a kinematic chain.

The subsystem of an industrial robot system that links the mechanical portion of the robot (manipulator) to the part being handled or worked on and gives the robot the ability to pick up and transfer parts and/or handle a multitude of differing tools to perform work on parts. It is commonly made up of four distinct elements: a method of attachment of the hand or tool to the robot tool mounting plate, power for actuation of tooling machines, mechanical linkages, and sensors integrated into the tooling. Examples include grippers, paint spraying nozzles, welding guns, and laser gauging devices.

End-Effector, Turret[2]: A number of end-effectors, usually small, that are mounted on a turret for quick automatic change of end-effectors during operation.

Endpoint Control[2]: Control wherein the motions of the axes are such that the endpoint moves along a pre-specified type of path line (straight line, circle, etc.)

Endpoint Rigidity[2]: The resistance of the hand, tool, or endpoint of a manipulator arm to motion under applied force.

Error-Absorbing Tool[2]: A type of robot end-effector able to compensate for small variations in position and orientation. Especially suitable for assembly tasks, where the insertion of components demands tight tolerance positioning and orientation of the parts. (See also Remote Center Compliance device).

Extended jaw[1]: An (optional) additional jaw situated at the end of an impactive gripper finger. It may, in preference to the finger itself, be modified to fit the profile of the object and it may be replaceable.

External Sensor[2]: A feedback device that is outside the inherent makeup of a robot system, or a device used to effect the actions of a robot system that are used to source a signal independent of the robot's internal design.

Fixture[2]: A device used for holding and positioning a workpiece without guiding the tool.

Flexibility (Gripper)[2]: The ability of a gripper to conform to parts that have irregular shapes and adapt to parts that are inaccurately oriented with respect to the gripper.

Flexible Fixturing[2]: Fixture systems with the ability of accommodating several part types for the same type of operation. The fixture can be robotic and change automatically according to sensor input detecting the part change.

Flexible Fixturing Robots[2]: Robots working in parallel, designed to hold and position parts on which other robots or people or automation can work.

Force Control[2]: A method of error detection in which the force exerted on the end-effector is sensed and fed back to the controller, usually by mechanical, hydraulic or electric transducers.

Force-Torque Sensors[2]: The sensors that measure the amount of force and torque exerted by the mechanical hand along three hand-referenced orthogonal directions and applied around a point ahead and away from the sensors.

Geometric Dexterity[2]: The ability of the robot to achieve a wide range of orientations of the hand with the tool center point in a specified position.

Grasp Planning[2]: A capability of a robot programming language to determine where to grasp object in order to avoid collisions during grasping or moving. The grasp configuration is chosen so that objects are stable in the gripper.

Gripper[1] [2]: The generic term for all prehension devices whether robotic or otherwise. Loosely defined in four categories: Impactive, Astrictive, Ingressive and Contigutive.
   The grasping hand of the robot which manipulated objects and tools to fulfill a given task.

Gripper axis[1]: A frame with its origin in the TCP(Tool Center Point). This coordinate system is used to specify the gripper orientation.

Gripper Changing System[1]: A module for rapid manual, but in most cases automatic, exchange of an end-effector using a standard mechanical interface. In doing so, all power and control cables must be disconnected and reconnected.

Gripper Design Factors[2]: Factors considered during the design of a gripper in order to prevent serious damage to the tool or facilitate quick repair and alignment. The factors include: parts' or tools' shape, dimension, weight, and material; adjustment for realignment in the x and y direction; easy-to-remove fingers; mechanical fusing (shear pins, etc.); locating surface at the gripper-arm interface; spring loading in the z(vertical) direction; and specification of spare gripper fingers.

Gripper External[2]: a type of mechanical gripper used to grasp the exterior surface of an object with closed fingers.

Gripper, Internal[2]: A type of mechanical gripper used to grip the internal surface of an object with open fingers.

Gripper Finger[1]: Rigid, elastic, or multi-link grasping organ to enclose or clasp the object to be handled.

Gripper Hand (Hand Unit)[1]: Grippers with multiple jointed fingers, each of them representing an open kinematic chain and possessing a high degree of freedom with f joints.

Gripper jaw[1]: The part of the gripper to which the fingers are normally attached. The jaw does not necessarily come into contact with the object to be gripped. Note: in some cases gripper fingers may be fitted with an additional small (extended) jaw at their ends.

Gripper, Soft[2]: A type of mechanical gripper which provides the capability of conforming to part of the periphery of an object of any shape.

Gripper, Swing Type[2]: A type of mechanical gripper which can move its fingers in a swinging motion.

Gripper, Translational[2]: a type of mechanical gripper which can move its own fingers, keeping them parallel.

Gripper, Universal[2]: A gripper capable of handling and manipulating many different object of varying weights, shapes, and materials.

Gripping Area[1]: The area of the prehension (gripper jaw) across which force is transmitted to the object surface. The larger the contact surface area of an impactive gripper, the smaller the pressure on the object surface.

Gripping Surface(s)[1] [2]: The passive contact surface between object and gripper, i.e., the surface which is subjected to prehension forces.

The surfaces, such as the inside of the fingers, on the robot gripper or hand that are used for grasping.

Hand (Robot's) [2]: A fingered gripper sometimes distinguished from a regular gripper by having more than three fingers and more dexterous finger motions resembling those of the human hand.

Hand Coordinate System [2]: A robot coordinate system based on the last axis of the robot manipulator.

Handchanger [2]: A mechanism analogous to a tool changer on a machining center or other machine tool. It permits a single robot arm to equip itself with a series of task-specific hands or grippers.

Hard Tooling [2]: Traditional tooling where every part to be processed in the robotic cell has its own fixtures and tools. It results in increased changeover time and processing delays.

Holding system [1]: A term often used for an active prehension system including a gripper, jaws and fingers. It may also be apply to a passive temporary retaining device.

Impactive gripper [1]: A mechanical gripper whereby prehension is achieved by impactive forces, i.e. forces which impact against the surface of the object to be acquired.

Ingressive gripper [1]: Ingression revers to the permeation of an objects surface by the prehension means. Ingression can be intrusive (pins) or non intrusive (hook and loop)

Inspection (Robotic) [2]: Robot manipulation and sensory feedback to check the compliance of a part or assembly with specifications. In such applications robots are used in conjunction with sensors, such as a video camera, laser, or ultrasonic detector, to check part locations, identify defects, or recognize parts for sorting.

Jamming [2]: In part assembly, jamming is a condition where forces applied to the part for part mating point in the wrong direction. As a result, the part to be inserted will not move.

Joint Geometry Information [2]: Geometry data for the mating of the parts to be joined, assembled, or welded.

Kinematic Chain [2]: *Pertaining to manipulator.* The combination of rotary and/or translational joints, or axes of motion.

Kinematic System [1]: *Pertaining to end-effector.* Mechanical unit (gear) converting drive motion of the prime mover into prehension action (jaw motion) with characteristic transmission rates for velocities and forces. The most often used kinematic components are lever, screw, and toggle lever gears. The gear determines the final velocity of the jaw movement, and the gripping force characteristics. Grippers without moving elements require no kinematics.

Magnetoadhesion [1]: Prehension force by means of a magnetic field (permanent or electrically generated).

Main Reference [2]: A geometric reference which must be maintained throughout a production process. The compliance with the references of the component elements of a subassembly guarantees the geometry of the complete assembly.

Manipulation (Robotic) [2]: The handling of objects, by moving, inserting, orienting, twisting, and so on, to be in the proper position for machining, assembling, or some other operation. In many cases it is the tool that is being manipulated rather than the object being processed.

Material Handling (Robotic) [2]: the use of the robot's basic capability to transport objects. It is common to find robots performing material-handling tasks and interfacing with other material-handling equipment such as containers, conveyors, guided vehicles, monorails, automated storage/retrieval systems, and carousels.

Mechanical Grip Devices [2]: The most widely used type of end-of-arm tooling in parts-handling applications. Pneumatic, hydraulic, or electrical actuators are used to generate a holding force which is transferred to the part via linkages and fingers. Some devices are able to sense and vary the grip force and grip opening.

Minimal Precedence Constraint (MPC) Method [2]: A method for the generation of assembly sequences based on the identification of geometric precedence constraints that implicitly represent all geometrically feasible assembly sequences. The minimal precedence constraint for an assembly component is defined as the alternative assembly states that will prevent the assembly of this component.

Mounting Plate [2]: The means of attaching end-of-arm tooling to an industrial robot. It is located at the end of the last axis of motion on the robot. The mounting plate is sometimes used with an adapter plate to enable the use of a wide range of tools and tool power sources.

Multi-gripper System [2]: A robot system with several grippers mounted on a turret-like wrist, or capable of automatically exchanging its gripper with alternative grippers, or having a gripper for multiple parts. A type of mechanical gripper enabling effective simultaneous execution of two or more different jobs effectively.

Multiple grippers [1]: Several grippers mounted on the same substrate, intended for the simultaneous prehension of more than two objects.

Multi-hand Robot Systems [2]: A class of robotic manipulators with more than one end-effector, enabling effective simultaneous execution of two or more different jobs. Design methods for each individual hand in a multi-hand system are similar to those of single hands, but must also consider the other hands

Noncontact Sensor [2]: A type of sensor, including proximity and vision sensors, that functions without any direct contact with objects.

Orientation [2]: Also known as positioning. The consistent movement or manipulation of an object into a controlled position and attitude in space.

Orientation Finding [2]: The use of a vision system to locate objects so they can be grasped by the manipulator or mated with other parts.

Palletizing/Depalletizing [2]: A term used for loading/unloading a carton, container, or pallet with parts in organized rows and possibly in multiple layers.

Part Mating [2]: The action of assembling two parts together according to the assembly design specifications. It occurs in four stages: approach, chamfer crossing, one-point contact, and two-point contact.

Part Mating Theory [2]: Predicts the success or failure mode of the assembly of common geometry parts, such as round pegs and holes, screw threads, gears, and some simple prismatic part shapes. Two common failure modes during two-point contact are wedging and jamming.

Payload [2]: The maximum weight that a robot can handle satisfactorily during its normal operations and extensions.

Peripheral Equipment[2]: The equipment used in conjunction with the robot for a complete robotic system. This equipment includes grippers, conveyors, part positioners, and part or material feeders that are needed with the robot.

Photoelectric Sensors[2]: A register control using a light source, one or more phototubes, a suitable optical system, and amplifier, and a relay to actuate control equipment when a change occurs in the amount of light reflected from a moving surface due to register marks, dark areas of a design, or surface defects.

Pick-and-Place[2]: A grasp-and-release task, usually involving a positioning task.

Pneumatic Pickup Device[2]: The end-of-arm tooling such as vacuum cups, pressurized bladders, and pressurized fingers.

Pose[2]: The robot's joints position for a particular end-effector position and orientation within the robot's workspace. Specific positions are named according to the tasks the robot is performing; for example, the home pose, which indicates the resting position of the robot's arm.

Position Control[2]: a control by a system in which the input command is the desired position of a body.

Position Finding[2]: The use of a vision system to locate objects so they can be grasped by a manipulator or mated with other parts.

Positioners[2]: Also known as positioning table, positioners are fixture devices for locating the parts to be processed n the required position and orientation. Positioners can be implemented as hard tooling devices or reprogrammable robotic devices which reduce the setup time and part changeover times. For instance, positioners are used in robotic arc welding to hold and position pieces to be welded. The movable axes of the positioner are sometimes considered additional robot axes. The robot controller controls all axes in order to present the seam to be welded by the robot's torch in the location and orientation taught or modified by adaptive feedback, or changes inserted by the operator, dynamically during execution.

Precision (Robot)[2]: A general concept reflecting the robot's accuracy, repeatability, and resolution.

Prehendability[1]: The suitability of an object to be automatically gripped. Dependant on the surface properties, weight and strength when exposed to prehension forces. This property can sometimes be enhanced by applying such surfaces or elements (handling adapters) which are required only for a particular purpose.

Prehension[1]: The act of acquiring and object in or onto the gripper. (we must modify some of these terms based on our chosen vocabulary (gripper/end-effector and object/workpiece)

Prehension planning[1]: Deals with the problem of how to ensure stable mating between robot gripper and workpiece. A prehension strategy must be chosen in such a way that it can be accomplished in a stable manner and collision free. Post prehension misalignment of the object is undesirable. In many circumstances, special constraints must be observed in order to avoid contact with certain parts of the object (forbidden zones)

Prehension systems[1]: Complete systems including grippers supplemented with additional units (subsystems), e.g., rotation, pivot and short-travel units, changing systems, joining (adjustment) tools, collision and overload protection mechanisms, measuring devices and other sensors.

Pressurized Bladder[2]: A pneumatic pickup device which is generally designed especially to conform to the shape of the part. The deflated bladder is placed in or around the part. Pressurized air causes the bladder to expand, contact the part, and conform to the surface of the part, applying equal pressure to all points of the contacted surface.

Pressurized Fingers[2]: A pneumatic pickup device that has one straight half, which contacts the part to be handled, one ribbed half, and a cavity for pressurized air between the two halves. Air pressure filling the cavity causes the ribbed half to expand and "wrap" the straight side around a part.

Prosthetic Robot[2]: A programmable manipulator or device that substitutes for lost functions of human limbs.

Protection system[1]: These are elements attached to the inner or outer part of the gripper which are activated in case of overload or collision in order to protect the robot and gripper from damage (warning signal, emergency stop activation, passive or active evasive movement).

Proximity Sensor[2]: A device which senses that an object is only a short distance away and /or measures how far away it is. Proximity sensors typically work on the principles of triangulation of reflected light, elapsed time for the reflected sound, intensity-induced eddy currents, magnetic fields, back pressure from air jets, and others.

Repeatability[2]: The envelope of variance of the robot tool point position for repeated cycles under the same conditions. It is obtained from the deviation between the positions and orientations reached at the end of several similar cycles.

Resolution[2]: The smallest incremental motion which can be produced by the manipulator. It serves as one indication of the manipulator accuracy. Three factors determining the resolution: mechanical resolution, control resolution, and programming resolution.

Retention[1]: Pertains to the post prehension status of an object already held in the gripper. Note: prehension and retention forces are not always equal.

Retro-reflective Sensing[2]: A photoelectric source consolidation method based on the aiming of the light beam into a white retro target feeding a photoelectric sensor.

Rigidity[2]: The property of a robot to retain its stiffness under loading and movement. Rigidity can be improved by features such as a cast-iron base, precision ball screws on all axial drives, ground and hardened spiral bevel gears in the wrist, brakes on the least stiff axes, and end-effector design that permits a workpiece or tool to be held snugly.

Robot Task[2]: Specification of the goals for the positioning of the object being manipulated by the robot, ignoring the motions required by the robot to achieve these tasks.

Robotic Assembly[2]: the combination of robots, people, and other technologies for the purpose of assembly in a technologically and economically feasible manner. Robotic assembly offers an alternative with some of the flexibility of people and the uniform performance of the fixed automation.

Robotic Fixturing[2]: a programmable fixture system that can accommodate a set of parts for processing in the same workcell.

Search Routine[2]: A robot function that searches for a precise location when it is not known exactly. An axis or axes move slowly in one direction until terminated by an external signal. It is used in stacking and unstacking of parts, locating workpieces, or inserting parts in holes.

Secondary References[2]: Geometric references used in assembling a main assembly component. The compliance with the references of the components guarantees the correct geometry of the completed assembly.

Selective Compliance Assembly Robotic Arm (SCARA)[2]: A horizontal-revolute configuration robot designed at Japan's Yamanachi University. The tabletop-size arm with permanently tilted, high-stiffness links sweeps across a fixtured area and is especially suited for small-parts insertion tasks in the vertical (z) direction.

Sensing[2]: The feedback from the environment of the robot which enables the robot to react to its environment. Sensory inputs may come from a variety of sensor types, including proximity switches, force sensors, and machine vision systems.

Sensor Coordinate System[2]: A coordinate system mounted over the workspace of the robot and assigned to a sensor.

Sensor Fusion (Sensor Integration)[2]: The coordination and integration of data from diverse sources to produce a usable perspective for a robotics system. A large number of sensors can be applied, and the information they gather from the work environment or workpiece is analyzed and integrated in a unique meaningful stream of feedback date to the robotic manipulator.

Sensor System[1][2]: Sensors pertinent to the task or prehension. This may include sensors built into the end-effector, possibly with integrated data pre-processing, for position detection, registration of object approach, determination of gripping force, path and angle measurements, slippage detection etc.
 The components of a robot system which monitor and interpret events in the environment. Internal measurement devices, also considered sensors, are part of closed axis-control loops and monitor joint position, velocity, acceleration, wrist force, and gripper force. External sensors update the robot model and are used for approximation, touch, geometry, vision, and safety. A data acquisition system uses data from sensors t o generate patterns. A data processing system then identifies the patterns and generates frames for the dynamic world-model processor.

Sensor Glove[2]: A robotics sensor capable of precision measurement of human gestures, with applications in surgery and telerobotics.

Sensory-Controlled Robot[2]: Also known as intelligent robot. A robot whose program sequence can be modified as a function of information sensed from its environment. The robot can be served or non-servoed.

Slip Sensors[2]: Sensors that measure the distribution and amount of contact area pressure between hand and objects positioned tangentially to the hand. They may be single-point, multiple-point (array), simple binary (yes-no), or proportional sensors.

Sorting (Robotic)[2]: The integrated operation of a sensor system and a robot for the discrimination of two or more types of workpieces.

Sucker[1]: Normally refers to a passive suction element (disk, cap or cup) which does not require active vacuum suction but relies on the evaluation of air by distortion of the element against the object surface.

Suction head[1]: A form of astrictive gripper which may consist of one or more vacuum suction elements (discs, caps or cups) from which air is actively evacuated by means of externally generated negative pressure.

Synchronization[1]: very specific, not sure if we need it.

Tactile Sensing[2]: The detection by a robot through contact by touch, force, pattern slip, and movement. Tactile sensing allows for the determination of local shape, orientation and feedback forces of a grasped workpiece.

Thermoadhesion[1]: Contigutive prehension force by means of thermal effects (usually in the form of freezing or melting).

Tool Center Point (TCP) [1][2]: Working pint at the end of a kinematic chain. The TCP serves also as a programmed reference point for an end effector and as a rule determines the origin of the tool frame. A coordinate system whose origin coincides with the TCP is called the tool frame. Multiple gripper heads may possess several TCPs or one main TCP with the rest being defined relative to the main TCP by tool offsets. A tool-related reference point that lies along the last wrist axis at a user-specified distance from the wrist.

Tool Coordinate System[2]: A coordinate system assigned to the end-effector.

Tool-Coordinate programming[2]: Programming the motion of each robot axis so that the tool held by the root gripper is always geld normal to the work surface.

Torque Control[2]: A method to control the motions of a robot driven by electric motors. The torque produced by the motor is treated as an input to the robot joint. The torque value is controlled by the motor current.

Torque/Force Controller[2]: A control system capable of sensing forces and torques encountered during assembly or movement of objects, and /or generating forces on joint torques by the manipulator which are controlled to reach desired levels.

Touch Sensors[2]: Sensors that measure the distribution and amount of contact area pressure between hand and objects perpendicular to the hand. Touch sensors may be single point, multiple-point (array), simple binary (yes-no), or proportional sensors, or may appear in the form of artificial skin.

Tracking[2]: A continuous position-control response to continuously changing input requirements.

Tracking (Line) [2]: The ability of a robot to work with continuously moving production lines and conveyors. Moving-base line tracking and stationary-base line tracking are the two methods of line tracking.

Tracking Sensor[2]: Sensors used by the robot to continuously adjust the robot path in real time while it is moving.

Vacuum Cups[2]: A type of pneumatic pickup device which attaches to parts being transferred via a suction of vacuum pressure created by a venturi or a vacuum pump.

Vision System[2]: A camera (or cameras) system interfaced to guide a robot to locate a part, identify it, direct the gripper to a suitable grasping position, pickup the part, and bring it to the work area. A coordinate transformation between the cameras and the robot must be carried out to enable proper operation of the system.

Wedging[2]: In rigid part assembly, a condition where two-point contact occurs too early in part mating, leading to the part that is supposed to be inserted appearing to be stuck in the hole. Unlike jamming, wedging is caused by geometric rather than ill-proportioned applied forces.

Workpiece or object[1]: A general term which refers to the component of object to be prehended or which is already under prehension by the gripper.

Wrist[2]: A set joints, usually rotational, between the arm and the hand or end-effector, which allow the hand or end-effector to be oriented relative to the workpiece.

Wrist force Sensor[2]: A structure with some compliant sections and transducers that serve as force sensors by measuring the deflections of the compliant sections. The types of transducers used are strain-gauge, piezoelectric, magnetostrictive, and magnetic.

b. Industrial Robot System Standards

This section includes a list of standards from the normative and bibliography references listed in ISO 10218; ANSI/RIA R15.06 and [Moon, 2009].

ISO 4413, Hydraulic fluid power — General rules relating to systems
ISO 4414, Pneumatic fluid power — General rules relating to systems
ISO/IEC Guide 51, Safety aspects — Guidelines for their inclusion in standards
ISO 7000, Graphical symbols for use on equipment — Index and synopsis
ISO 8373:1994, Manipulating industrial robots — Vocabulary
ISO 9283:1998, Manipulating industrial robots — Performance criteria and related test methods
ISO 9409 (all parts), Manipulating industrial robots — Mechanical interfaces
ISO 9946, Manipulating industrial robots — Presentation of characteristics
ISO 10218-1, Robots and robotic devices — Safety requirements — Part 1: Industrial robot
ISO 10218-2, Robots and robotic devices — Safety requirements — Part 2: Industrial robot system and integration
ISO 11161, Safety of machinery — Industrial automation systems — Safety of integrated manufacturing systems — Basic requirements
ISO 11593:1996 Manipulating industrial robots --Automatic end effector exchange systems -- Vocabulary/presentation of characteristics
ISO 12100:2010 Safety of machinery -- General principles for design -- Risk assessment and risk reduction
ISO/TR 13309:1995 Manipulating industrial robots -- Informative guide on test equipment and metrology methods of operation for robot performance evaluation in accordance with ISO 9283
ISO 13849-1:2006, Safety of machinery — Safety-related parts of control systems — Part 1: General principles for design
ISO 13850, Safety of machinery — Emergency stop — Principle for design
ISO 13851, Safety of machinery — Two-hand control devices — Functional aspects and design principles
ISO 13854, Safety of machinery — Minimum gaps to avoid crushing of parts of the human body
ISO 13855, Safety of machinery — Position of protective equipment with respect to the approach speeds of parts of the human body
ISO 13856-1, Safety of machinery — Pressure-sensing protective devices — Part 1: General principles for design and testing of pressure-sensitive mats and pressure-sensitive floors
ISO 13857, Safety of machinery — Safety distances to prevent danger zones being reached by the upper limbs and lower limbs
ISO 14118, Safety of machinery — Prevention of unexpected start-up
ISO 14119, Safety of machinery — Interlocking devices associated with guards — Principles for design and selection
ISO 14120, Safety of machinery — Guards — General requirements for the design and construction of fixed and movable guards
ISO/TR 11688-1:1995 Acoustics — Recommended practice for the design of low-noise machinery and equipment — Part 1: Planning
ISO 14123, Safety of machinery - Reduction of risks to health from hazardous substances emitted by

machinery
ISO 14159 Safety of machinery - Hygiene requirements for the design of machinery
ISO 14539:2000 Manipulating industrial robots --Object handling with grasp-type grippers -- Vocabulary and presentation of characteristics
ISO/TS 15066 - Technical specification on collaborative workspace (under elaboration)
ISO 19353, Safety of machinery - Fire prevention and protection
IEC 60204-1:2005, Safety of machinery — Electrical equipment of machines — Part 1: General requirements
IEC 60364-7-729, Low-voltage electrical installations — Part 7-729: Requirements for special installations or locations — Operating or maintenance gangways
IEC 61000-6-2, Electromagnetic compatibility (EMC) — Part 6-2: Generic standards — Immunity for industrial environments
IEC 61000-6-4, Electromagnetic compatibility (EMC) — Part 6: Generic standards — Section 4: Emission standard for industrial environments
IEC 61496-1, Safety of machinery — Electro-sensitive protective equipment
IEC 61496-2, Safety of machinery — Electro-sensitive protective equipment — Part 2: Particular requirements for equipment using active opto-electronic protective devices (AOPDs)
IEC 61800-5-2 Adjustable Speed Electrical Power Drive Systems
IEC 62061:2005, Safety of machinery — Functional safety of safety-related electrical, electronic and programmable control systems
ISO/CIE 8995-1, Lighting of work places — Part 1: Indoor
EN 563, Safety of machinery - Temperatures of touchable surfaces - Ergonomics data to establish temperature limit values for hot surfaces
EN 1093, Safety of machinery - Evaluation of the emission of airborne hazardous substances
EN 1127, Explosive atmospheres - Explosion prevention and protection
EN 12198, Safety of machinery - Assessment and reduction of risks arising from radiation emitted by machinery
CEN/TR 14715, Safety of machinery - Ionizing radiation emitted by machinery - Guidance for the application of technical standards in the design of machinery in order to comply with legislative requirements
BGIA/DGUV study - Procedural Guideline for the Arrangement of Workplaces with Collaborative Robots
NFPA 70E - 2009 - Standard for Electrical Safety in the Workplace – revised to address safety gaps and increase electrical worker protection from Shock, electrocution, arc flash, and arc blast.
ANSI/NFPA 79-1997 - Electrical Standard for Industrial machinery
ANSI/UL 1740-1998 - Safety Standard for robots and robotic equipment
OSHA 1904 - General requirement for recording and reporting occupational injuries and illnesses
OSHA 1910.147-Control of hazardous energy (Lockout/Tagout)
OSHA 1910.212 - General requirements for all machines (Machine guarding)
OSHA 1910.219 - Mechanical power transmission apparatus
ANSI B11.19-1990 (R1996), Safeguarding performance criteria
ANSI B11.20-1991 (R1996), Safety requirements for flexible manufacturing systems/cells
ANSI Z49.1 - 1994, Safety in welding, cutting and allied processes
ANSI Z136.1-1993, Safe use of lasers
ANSI Z244.1-1982 (R1993), Safety Requirements for the Lock Out/Tag Out of Energy Sources
ANSI Z535.1 - 1998, Safety Color Code
ANSI Z535.2-1998, Environmental and Facility Safety Signs
ANSI Z535.3-1998, Criteria for Safety Symbols and Labels
ANSI Z535.4-1998, Product Safety Signs and Labels
ANSI Z535.5-1998, Accident Prevention Tags (for Temporary Hazards)
ANSI/ASME B15.1-1992, Safety Standards for Mechanical Power Transmission Apparatus

ANSI/AWS D16.2-1994, Components of Robotic and Automatic Welding
ANSI/UL508-1988, Industrial Control Equipment
ANSI/UL969-1991, Standard for safety-marking and labeling systems
UL 991 - Tests for safety-related controls employing solid-state devices
UL 1998-Safety-related software

c. Design for Assembly – Boothroyd-Dewhurst Method

The Boothroyd-Dewhurst Design for Assembly (DfA) evaluation [AMI, 2011] centers on establishing the cost of handling and inserting component parts. The process (see Figure 24) can be applied to manual or automated assembly, which is further subdivided into high-speed dedicated or robotic. An aid to the selection of the assembly system is also provided by a simple analysis of the expected production volume, payback period required, number of parts in the assembly, and number of product styles.

Regardless of the assembly system, parts in the assembly are evaluated in terms of ease of handling and ease of insertion, and a decision is made as to the necessity of the part in question. The findings are then compared to synthetic data, and from this a time and cost are generated for the assembly of that part. The opportunity for reducing this is found by examining each part in turn and identifying whether each exists as a separate part for fundamental reasons. These fundamental reasons are:

- During operation of the product, does the part move relative to all other parts already assembled?
- Must the part be of a different material from all other parts already assembled? Or isolated from them?
- Must the part be separate from all those already assembled because otherwise necessary assembly or disassembly of other separate parts would be impossible?

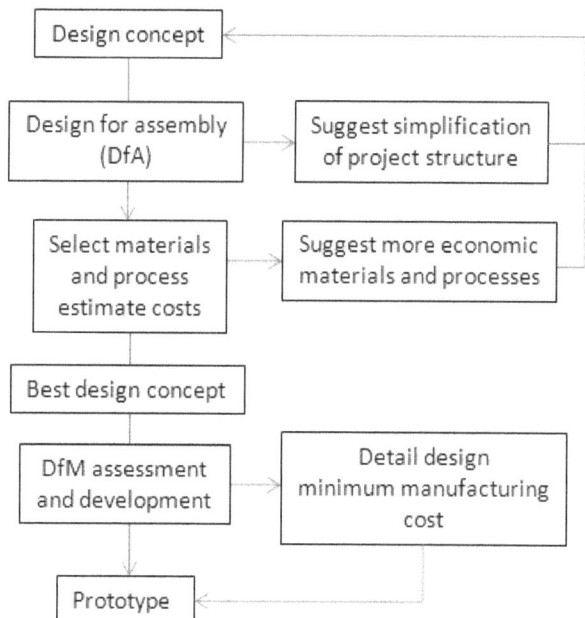

Figure 24 - Boothroyd-Dewhurst Design for Assembly Method

The second stage of the analysis is to examine the handling and insertion of each component part. For manual assembly, a two-digit handling code and a two-digit insertion code are identified from synthetic data tables. The tables categorize components with respect to their features for handling such as size, weight, and required amount of orientation. For insertion, there are categories for part alignment, the type of securing method, and whether the part is secured on insertion or as a separate process. These codes are then cross-referenced to identify the time for that operation from the table.

The codes and subsequent times are used to determine a number of metrics:

- **Assembly time (TM)** is determined by summing the handling and insertion times
- **Assembly cost (CM)** is proportional to TM by a factor that accounts for wage rate and overheads
- **Theoretical minimum number of parts (NM)** is the summation of all those essential parts determined during the first stage
- **Design efficiency** is defined as the ideal assembly time divided by the estimated assembly time
- The **ideal assembly time** is given by 3NM, where the 3 represents a handling time of 1.5 seconds and insertion time of 1.5 seconds, for an ideal component.
- The **estimated assembly time** is TM.

Though costs and times are determined, care must be taken in the use of these values in an absolute sense. As with other techniques, values are best used for comparing redesigns.

d. Types of Force Control Functions

Table 5 lists types of force control functions and descriptions from the [Fanuc, 2007] force control sensor operator's manual. Also, Figure 25 shows drawings of example assembly operations to better acquaint the reader with the types of assembly operations that can be accomplished using robotic force control.

Table 5 - Types of Force Control Functions
[copyright information-use permission granted by Fanuc Robots]

Function	Description
Unused	Schedule data is not used. Force control cannot be performed using an Unused schedule. See 3.5.1 "Unused".
Constant Push	This function is used to gently bring the robot hand into contact with the workpiece, for instance, for contact evaluation, temporary placement, and the arrangement of components along a particular guide. See 3.5.2 "Constant Push / Face Match"..
Face Match	This function is used to match the face of the workpiece held by the robot hand with the face of the object, such as when inserting a workpiece into the chuck of a machine tool. See 3.5.2 "Constant Push / Face Match".
Bearing Insert	This function inserts a workpiece in the same way as a main bearing is inserted into a car engine. See 3.5.3 "Shaft Insert / Bearing Insert / Groove Insert / Square Insert".
Shaft Insert	This function inserts a cylindrical mechanical component such as a shaft or a positioning pin. See 3.5.3 "Shaft Insert / Bearing Insert / Groove Insert / Square Insert".
Phase Match Ins.	This function first performs phase matching then performs insertion, as insertion with key and as spline insertion in a car transmission. This function is similar to "Phase Search". The number of parameters to be set for this function is smaller than that for "Phase Search", and consequently this function provides limited capabilities. See 3.5.4 "Phase Match Insert".
Groove Insert	This function inserts a quadrangular prism workpiece into a groove. See 3.5.3 "Shaft Insert / Bearing Insert / Groove Insert / Square Insert".
Clutch Insert	This function is used to assemble a clutch for the automatic transmission of an automobile. The function performs phase matching around the insertion axis and searches for a position on a plane perpendicular to the insertion axis at the same time. See also the description of the "Clutch Search" function. See 3.5.5 "Clutch Insert".
Search	This function absorbs the initial position and attitude errors that are present before the start of force control. Errors can be absorbed in the five directions (two translation directions plus three rotation directions) except the insertion direction. See 3.5.6 "Search Function".
Phase Search	This function performs phase matching of teeth as in a case of a key shaft and gear engagement. See 3.5.6 "Search Function". This function is similar to the "Phase Match Ins." function but differs in the following: - When the torque is sensed during phase matching, "Phase Search" causes an inversion in the search direction. - "Phase Search" performs phase matching by slightly changing the rotation velocity and torque so as not to damage the workpiece. - While "Phase Match Ins." performs phase matching then performs insertion, "Phase Search" performs phase matching only. To perform insertion successively after phase matching, execute "Shaft Insert" in succession. For successive execution, see Section 3.7, "SUCCESSIVE EXECUTION OF FORCE CONTROL INSTRUCTIONS".
Hole Search	This function performs a search operation on the plane perpendicular to the insertion direction. For shaft insertion, for example, the positioning error at the start of force control needs to be within the chamfer amount. However, the hole search function enables insertion even when there is a positioning error larger than the chamfer amount. For the successive execution, See Section 3.7 "SUCCESSIVE EXECUTION OF FORCE CONTROL INSTRUCTIONS", See 3.5.6 "Search Function".
Clutch Search	This function is used to assemble a clutch for the automatic transmission of an automobile. The function performs phase matching around the insertion axis and searches for a position on a plane perpendicular to the insertion axis at the same time. This function is similar to "Clutch Insert" but differs in the way for searching for a position on a plane. This function allows insertion when a larger initial positioning error than that permitted in "Clutch Insert" is present. See 3.5.6 "Search Function".
Square Insert	This function inserts a quadrangular prism workpiece into a rectangular hole. See 3.5.3 "Shaft Insert / Bearing Insert / Groove Insert / Square Insert".
Contouring	This function traces the surface of a workpiece with applying a specified force. Used with a sander, this function can perform grinding. See Subsection 3.5.7, "Contouring".
Contouring End	This instruction ends Contouring being executed. See Subsection 3.5.7, "Contouring".

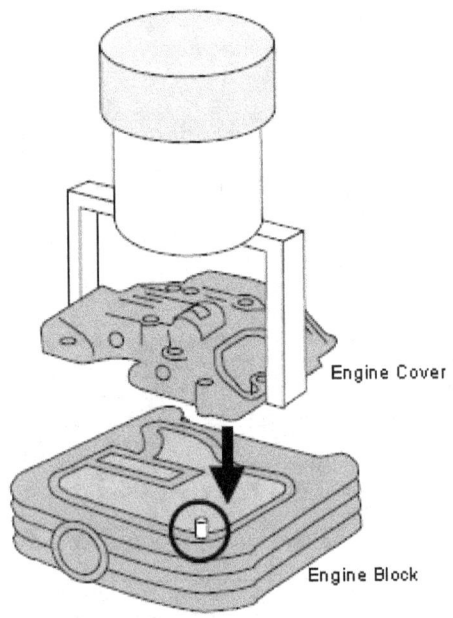

Engine Assembly with "Shaft Insert"
(Insertion of Knock Pin)

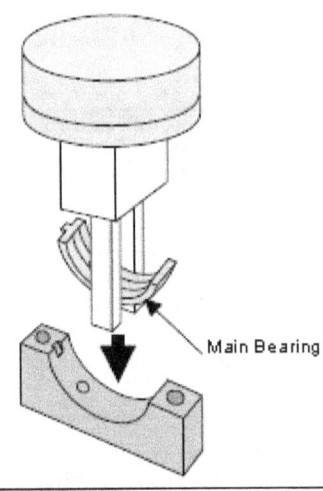

Engine Main Bearing Assembly with "Bearing Insert"

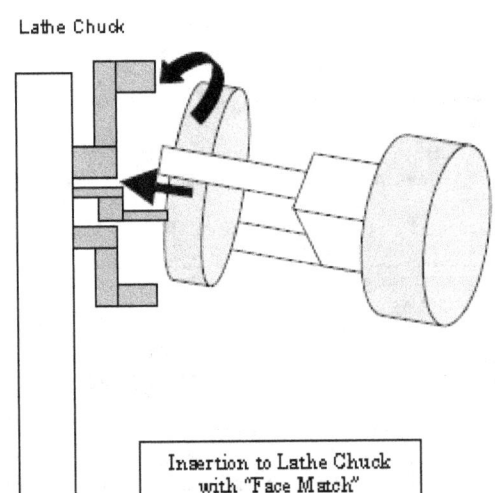

Insertion to Lathe Chuck with "Face Match"

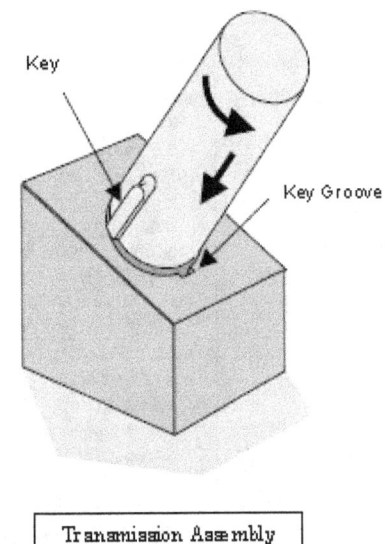

Transmission Assembly with "Phase Match Ins"

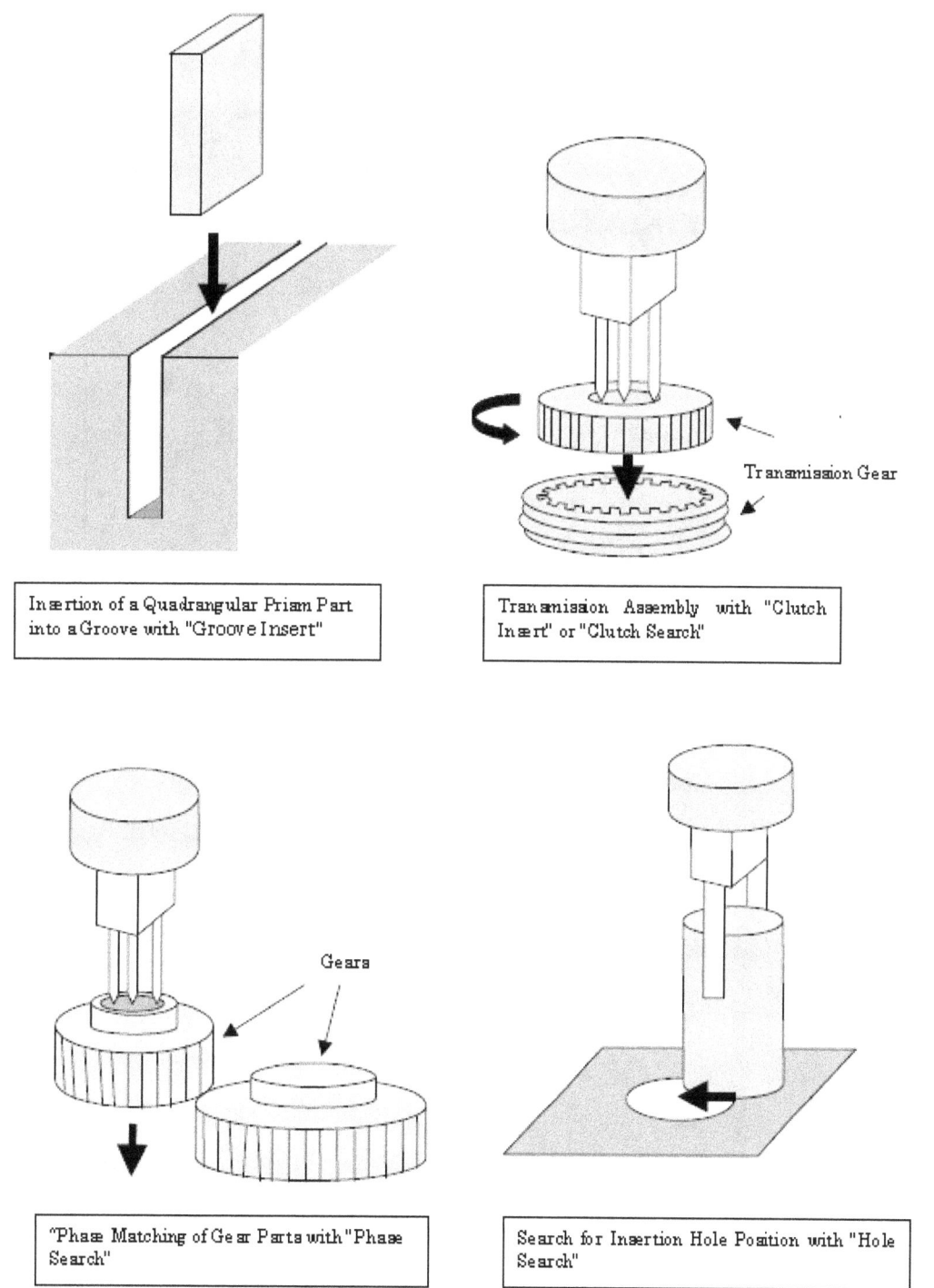

Figure 25 – Types of assembly functions currently accomplished using robotic force control. [Fanuc, 2007]
[copyright image/information-use permission granted by Fanuc Robots]

www.ingramcontent.com/pod-product-compliance
Lightning Source LLC
Chambersburg PA
CBHW081857170526
45167CB00007B/3047